AF832414

Studies in Space Policy

Volume 26

Series Editor

European Space Policy Institute, Vienna, Austria

Edited by: European Space Policy Institute, Vienna, Austria
Director: Jean-Jacques Tortora

Editorial Advisory Board:
Marek Banaszkiewicz
Karel Dobeš
Genevieve Fioraso
Stefania Giannini
Gerd Gruppe
Max Kowatsch
Sergio Marchisio
Fritz Merkle
Margit Mischkulnig
Dominique Tilmans
Frits von Meijenfeldt
https://espi.or.at/about-us/governing-bodies

The use of outer space is of growing strategic and technological relevance. The development of robotic exploration to distant planets and bodies across the solar system, as well as pioneering human space exploration in earth orbit and of the moon, paved the way for ambitious long-term space exploration. Today, space exploration goes far beyond a merely technological endeavour, as its further development will have a tremendous social, cultural and economic impact. Space activities are entering an era in which contributions of the humanities — history, philosophy, anthropology —, the arts, and the social sciences — political science, economics, law — will become crucial for the future of space exploration. Space policy thus will gain in visibility and relevance. The series Studies in Space Policy shall become the European reference compilation edited by the leading institute in the field, the European Space Policy Institute. It will contain both monographs and collections dealing with their subjects in a transdisciplinary way.

The volumes of the series are single-blind peer-reviewed.

More information about this series at http://www.springer.com/series/8167

Annette Froehlich
Editor

On-Orbit Servicing: Next Generation of Space Activities

ESPI
European Space Policy Institute

Springer

Editor
Annette Froehlich
European Space Policy Institute
Vienna, Austria

ISSN 1868-5307 ISSN 1868-5315 (electronic)
Studies in Space Policy
ISBN 978-3-030-51558-4 ISBN 978-3-030-51559-1 (eBook)
https://doi.org/10.1007/978-3-030-51559-1

© The Editor(s) (if applicable) and The Author(s), under exclusive license to Springer Nature Switzerland AG 2020
This work is subject to copyright. All rights are solely and exclusively licensed by the Publisher, whether the whole or part of the material is concerned, specifically the rights of translation, reprinting, reuse of illustrations, recitation, broadcasting, reproduction on microfilms or in any other physical way, and transmission or information storage and retrieval, electronic adaptation, computer software, or by similar or dissimilar methodology now known or hereafter developed.
The use of general descriptive names, registered names, trademarks, service marks, etc. in this publication does not imply, even in the absence of a specific statement, that such names are exempt from the relevant protective laws and regulations and therefore free for general use.
The publisher, the authors and the editors are safe to assume that the advice and information in this book are believed to be true and accurate at the date of publication. Neither the publisher nor the authors or the editors give a warranty, expressed or implied, with respect to the material contained herein or for any errors or omissions that may have been made. The publisher remains neutral with regard to jurisdictional claims in published maps and institutional affiliations.

This Springer imprint is published by the registered company Springer Nature Switzerland AG
The registered company address is: Gewerbestrasse 11, 6330 Cham, Switzerland

Preface

On-Orbit Servicing (OOS) is a clear sign that space activities have entered a new phase. The first phase was characterised by the emergence of the major spacefaring nations and includes the space race. During this era, space activities were overwhelmingly governmental or military in nature. The second phase saw the opening up of the sector to commercial actors, with activities no longer centered as heavily on governments or militaries. New activities in commercial fields such as telecommunications multiplied. The latest phase has witnessed the rise of private space actors and is characterised by the emergence of an in-space economy, with a much greater focus on new economic activities and opportunities in orbit. Emerging space actors from the Global South are also participating in this new phase to a much greater extent than hitherto. Also, whereas in previous phases the consequences of space activities were given scant attention, this focus is now at the forefront, and as such, OSS is at the heart of this new phase of space activities. A dedicated volume that brings together a range of new and diverse insights on OSS is thus timely. OSS itself is also evolving, with early operations carried out by astronauts or cosmonauts while further missions are likely to be conducted by robots. As such, in this volume, the implications of OSS will be explored from a variety of angles.

Space debris mitigation is a critical issue in terms of the sustainability of outer space activities and has been the subject of vigorous debate in the United Nations Committee on the Peaceful Uses of Outer Space (UNCOPUOS) and elsewhere. Various regulations and standards have consequently been created, including the UNCOPUOS Space Debris Mitigation Guidelines, the European Space Agency (ESA) Debris Mitigation Requirements, and the U.S. Orbital Debris Mitigation Standard Practices. While these regulations are welcome, OOS presents both a new challenge and opportunity, since it is opening up the possibility of in-space repairs, refuelling, and refurbishment of spacecraft. OOS is thus a further evolution of debris mitigation. The possibility of servicing satellites can play a critical role in extending the satellite lifecycle, and as such also supports the ESA's Clean Space initiative. However, it is necessary to approach the topic of OSS from more than just a technical feasibility angle, since it raises questions and concerns from political, legal, economic, and security perspectives.

Indeed, from a military standpoint, due to the dual-use nature of OOS, norms of behaviour must be formulated and implemented in terms of transparency and confidence-building measures (TCBMs) in outer space activities. Moreover, the principle expressed in the Outer Space Treaty that all space activities shall be for peaceful purposes still has no uniformly accepted definition, and it is vital that this issue is clarified so it can play a part in preventing hostile actions under the guise of OSS. As such, a distinction may have to be made between governmental and non-governmental OSS operators. This is only one of many legal questions and challenges arising from OSS. Others include, for example, the status of re-worked and recycled space objects, especially in terms of a potential transfer of ownership and jurisdiction. There is also the question of control in terms of operations performed in orbit, i.e. in the international sphere. A distinction will need to be made between a repair operation, and the use of recycled material from disused spacecraft to create new objects manufactured through, for example, 3D printing utilising the recycled material. This issue has serious legal implications regarding the status of the manufactured objects and applicable law, including liability and registration (for example, who will be considered as the launching state?). In turn, this will have an impact on insurance law and risk management. For instance, satellites previously considered total losses might be viewed differently by insurance companies in the context of OSS technology. Risk management thus has to be reviewed.

Finally, OSS also has implications for emerging space actors in the Global South. While their capabilities are more limited in terms of participating as OSS providers, this will directly affect their space operations. For example, OSS has a major transformative potential for emerging space actors who have experienced losses of expensive spacecraft providing critical services. For this reason, while OSS may seem to be a concern of the developed space actors, emerging space nations, particularly in Africa, will have to play a much more active role in the debates on the legal, political, economic, security, and risk issues if they are to safeguard their own interests. The critical challenge is to achieve a paradigm shift in perspective so that developing countries recognise the need and urgency of being involved in discussions around OSS, as opposed to leaving it up to the developed space actors.

Providing a novel approach to and perspectives on OSS this publication will be of great interest to practitioners, academics, and students working in the space sector and related fields.

Vienna, Austria
May 2020

Dr. Annette Froehlich
European Space Policy Institute (ESPI)
Seconded by German Aerospace Center (DLR)

Contents

1 **Legal Approach on the Dual-Use Nature of On-Orbit Servicing Programs** 1
Anne-Sophie Martin

2 **Risk Management and Insurance of On-Orbit Servicing** 13
Katarzyna Malinowska

3 **Legal Aspects of Space Recycling** 35
Zhuang Tian and Yangyang Cui

4 **On-Orbit Servicing: Security and Legal Aspects** 55
John Tziouras

5 **On-Orbit Servicing: Space for Africa?** 69
André Siebrits

6 **The Space Race on Sustainability: Business and Legal Challenges for On-Orbit-Servicing** 91
Claudiu Mihai Tăiatu

7 **On-Orbit Servicing, Opportunity and Liability** 123
Christoffel Kotze

8 **On-Orbit Servicing from a Legal and Policy Perspective** 141
Margaux Morssink

9 **Legal Aspects Relating to On-Orbit Servicing and Active Debris Removal** 159
Ewan Wright

Chapter 1
Legal Approach on the Dual-Use Nature of On-Orbit Servicing Programs

Anne-Sophie Martin

Abstract On-orbit servicing (OOS) activities consist in a focused action, through a '*space tug*' in order to maintain, repair, upgrade, refuel or de-orbit a spacecraft while it is in orbit. These activities require the servicer spacecraft to approach, *rendezvous* and interoperate with the space asset to another State, Agency or private company. In this context, even if OOS missions are as a first step not considered as military activities, the basic abilities of the system are of dual-use nature, allowing civil and military capacities. So, one can highlight the fact that an on-orbit service vehicle might represent a risk to the peaceful uses of outer space. Thus, one of the main legal challenges is to figure out how the OOS vehicle is used and the purpose of its mission. As a first step, the chapter analyses the article IV of the Outer Space Treaty. Then, it underscores that if these technologies cannot be considered as weapons; they can be viewed as a menace because of their dual-use nature. Hence, the chapter considers the criteria for *Transparency and Confidence Building Measures* in order to identify norms of behavior allowing to reduce the risks of misunderstanding that could induce crisis or conflict in outer space.

---

[1]R.S. Jakhu, J.N. Pelton (eds), *Global Space Governance: An International Study*, Springer, 2017, 331 ss; SpaceNews, *On-Orbit Satellite Servicing: The Next Big Thing in Space?*, November 17, 2017: https://spacenews.com/on-orbit-satellite-servicing-the-next-big-thing-in-space/; SpaceNews, *In-Orbit Services Poised to Become Big Business*, June 10, 2018: https://spacenews.com/in-orbit-services-poised-to-become-big-business/.

A.-S. Martin (✉)
Department of Political Sciences, Sapienza University of Rome, Rome, Italy
e-mail: annesophie.martin@uniroma1.it

© The Editor(s) (if applicable) and The Author(s), under exclusive license to Springer Nature Switzerland AG 2020
A. Froehlich (ed.), *On-Orbit Servicing: Next Generation of Space Activities*, Studies in Space Policy 26, https://doi.org/10.1007/978-3-030-51559-1_1

1.1 Introduction—Recent Practice in Space Servicing Missions

On-orbit servicing (OOS) and Rendezvous and Proximity (RPO) missions are becoming a reality.[1] They represent the next generation of space activities[2] with the purpose to approach and to operate space assets in orbit. Recently, the private company Northrop Grumman successfully launched an OOS mission.[3] Other companies such as Airbus Defence and Space,[4] MDA Corporation[5] are developing this technology. Moreover, space insurance companies are implementing new underwriting process based on these novel business plans.[6]

The European Space Agency is also enhancing this new technology with *Clean Space.*[7] The mission has various purposes, and in particular space debris removal.[8] It will allow to refueling satellites, repairing and moving them to new orbits.

These missions depict a new paradigm in space activities. Indeed, on one hand there is a requirement to prevent collision in orbit; on the other hand, in case of OOS/RPO, the issue is to dock two space assets, which raise some legal issues,[9] in particular to what extent these technologies can be viewed as peaceful. Furthermore, if at first sight OOS are not military space activities; nevertheless, the basic abilities of these systems are of dual use nature, enabling also military capabilities. Consequently, this technology has the capacity to being used to harm other space asset.

---

[2]ESPI Briefs, *In-Orbit Servicing: Challenges and Implications of an Emerging Capability*, n. 38, February 2020.

[3]The Verge, *Two commercial satellites just docked in space for the first time*, Feb. 26, 2020: https://www.theverge.com/2020/2/26/21154426/commercial-satellites-docking-space-northrop-grumman-intelsat; SpaceNews, *Northrop Grumman's MEV-1 servicer docks with Intelsat satellite*, Feb. 26, 2020: https://spacenews.com/northrop-grummans-mev-1-servicer-docks-with-intelsat-satellite/; *see also* EOportal Directory, MEV-1 (Mission Extension Vehicle-1): https://directory.eoportal.org/web/eoportal/satellite-missions/m/mev-1; Northrop Grumman, *Mission Extension Vehicle*: https://www.northropgrumman.com/space/space-logistics-services/mission-extension-vehicle/; SpaceNews, *Intelsat-910 Satellite, with MEV-1 Servicer Attached, Resumes Service*, April 17, 2020: https://spacenews.com/intelsat-901-satellite-with-mev-1-servicer-attached-resumes-service/; *see also* Space, *Ailing Intelsat Satellite Begins New Life in Orbit After Historic Servicing Mission Success*, April 17, 2020: https://www.space.com/ailing-intelsat-satellite-revived-by-mev-1-mission-success.html.

[4]Airbus website, *O. Cubed Services*: https://www.airbus.com/space/Services/on-orbit-services.html.

[5]MDA website, *Robotics and On-Orbit Servicing*: https://mdacorporation.com/isg/robotics-automation/space-based-robotics-solutions/.

[6]J.J. Klein, *Understanding Space Strategy—The Art of War in Space*, Routledge, 2019, 258p.

[7]ESA website, *In-Orbit Servicing: Mission profile*: http://www.esa.int/ESA_Multimedia/Images/2019/07/In-Orbit_Servicing_Mission_profile.

[8]Ibidem.

[9]A.S. Martin, S. Freeland, *Exploring the Legal Challenges of Future On-Orbit Servicing Missions and Proximity Operations*, Journal of Space Law, 43.2, 2019, 196–222.

Since the beginning of space age, States have used outer space for military activities and strategic operations.[10] Moreover, dual-use technologies have always been a feature of space technologies and missions such as in case of remote sensing, telecommunications and navigation.[11] It is necessary to recall that the legal regime of outer space is based on the principle of peaceful uses. However, there is a certain ambiguity on the meaning of "peaceful purpose".[12]

The concept of "peaceful purposes", also contained in the preamble to the Outer Space Treaty (OST),[13] has been interpreted by scholars as "non-aggressive purposes".[14] Although some States and commentators have suggested "peaceful purposes" to indicate "non-military purposes", this interpretation does not correspond with the practice of States to deploy military and dual-use satellites in orbit around the Earth.[15] The common idea is that "peaceful use" means "non-aggressive use".[16] By contrast, space activities can be conducted for peaceful purposes while also contributing to security and defense interests. The issue remains of fundamental importance due to the growing development of military activities, dual-use satellites and new threats in space. The military use of outer space is now consolidated both by consistent and numerous interpretations about the words used in Article IV, and by States practices.[17]

---

[10] M. Bourbonnière, *National-Security Law in Outer Space: The Interface of Exploration and Security*, 70 J. Air Law and Commerce (2005), 3–62; Jinyuan Su, *Use of Outer Space for Peaceful Purposes: Non-Militarization, Non-Aggression and Prevention of Weaponization*, 36 J. Space L. 253 (2010); Jackson Maogoto & Steven Freeland, *From Star Wars to Space Wars—The Next Strategic Frontier: Paradigms to Anchor Space Security*, 33 Annals Air & Space L. 10 (2008).

[11] *See generally* P. Gasparini Alves, *Evolving Trends in the Dual Use of Satellites*, UNIDIR, New York, 1996; F. Lyall, P.B. Larsen, Space Law A Treatise, Routledge, 2018, 448 ss.

[12] P.J. Blount, *Limits on Space Weapons: Incorporating the Law of War Into the Corpus Juris Spatialis*, Proceedings of the 51st Colloquium on the Law of Outer Space; P.J. Blount, *Developments in Space Security and Their Legal Implications*, Law/Technology, 44(2), 2011, 18–39; M. Bourbonnière, R.J. Lee, *Legality of the Deployment of Conventional Weapons in Earth Orbit: Balancing Space Law and the Law of Armed Conflict*, European Journal of International Law, 18(5), 2007, 873–901.

[13] *Treaty on Principles Governing the Activities of States in the Exploration and Use of Outer Space, including the Moon and Other Celestial Bodies*, October 10, 1967, 610 U.N.T.S. 205.

[14] C.Q. Christol, *The Modern International Law of Outer Space*, 1982, New York, Pergamon Press, 29–30; G. Catalano-Sgrosso, *Diritto internazionale dello spazio*, Firenze, LoGisma, 2011, 70; *see also* S. Aoki, *Law and Military Uses of Outer Space*, in R.S. Jakhu, P.S. Dempsey, *Routledge Handbook of Space Law*, Routledge, 2017, 197–224.

[15] R.J. Lee, *The Jus ad Bellum in Spatialis: the Exact Content and Practical Implications of the Law on the Use of Force in Outer Space*, 29 J. SpaceL. (2003) 93, 97–98.

[16] P.J. Blount, *Space Security Law*, Oxford Research Encyclopedia, Planetary Science, June 2018, 30p.

[17] S. Freeland, R.S. Jakhu, *The Applicability of the United Nations Space Treaties During Armed Conflict*, 2015 Proc. IISL, 11p; K.U. Schrogl, J. Neumann, *Article IV*, in S. Hobe, B. Schmidt-Tedd, K.U. Schrogl, CoCoSL, Vol. 1, Köln: Carl Heymanns Verlag, 2009, 70–93.

The military uses of outer space, with a growing dual-use technology tendency, are now widespread amongst numerous space-faring nations. Dual-use nature of space technologies means that they can be used for civil and military purposes, as well as they are funded by civil and military entities (ex: national space agency and ministry of defense).[18] States have always used satellites for military purposes, given their important strategic role to guarantee national security, or to support military or humanitarian operations on Earth. Indeed, warfare is henceforth unthinkable without the support of space capacities.[19]

Furthermore, the fact this technology is civil or commercial in nature does not lessen its strategic value.[20] Even if commercially developed, they have the potential to expand security concerns and could possibly pave the way to an arms race in outer space.[21]

The chapter examines the legality of these technologies in view of their dual use capacity. The first part analyses the Article IV of the Outer Space Treaty which is commonly regarded as the focal point dealing with the military uses of outer space (1). The second part underscores the efforts of the international community to address the risk of hostile actions in outer space due to OOS activities (2). The last part concerns the necessity to implement *transparency and confidence building measures* for on-orbit servicing missions (3).

1.2 Article IV of the OST in the Light of Military and Dual-Use Activities in Outer Space

As a first step, when addressing the military use of outer space, it is necessary to take into account Article III of the Outer Space Treaty because it determines that space law is associated with international law.[22] Moreover, the article makes reference to

---

[18] L. Bianchi, *Diritto spaziale e difesa: uso duale e security*, in A.F. Biagini, M. Bizzarri (a cura di) *Spazio. Scenari di collaborazione, note di diritto internazionale*, Firenze, Passigli Editori, 2013, 23.

[19] D. Blake, *Military Strategic Use of Outer Space*, in H. Nasu, R. McLaughlin (eds.), *New Technologies and the Law of Armed Conflict*, Netherlands, Asser Press, 2014, 97–114.

[20] J.N. Pelton, *Satellite Security and Performance in an Era of Dual Use*, Online Journal of Space Communication, Issue n°6, Security and Performance, Winter 2004.

[21] J.A. Lewis, *Reconsidering Deterrence for Space and Cyberspace*, in M. Krepon, J. Thompson (eds.), *Anti-Satellite Weapons, Deterrence and Sino-American Space Relations*, Washington, DC: Stimson Center, 2013, 61–80; M. Cervino, S. Corradini, S. Davolio, *Is the peaceful use of outer space being ruled out?*, Space policy 19, 2003, 231–237.

[22] M. Lachs, *The Law of Outer Space*, Leiden, 1972, p. 21 ss; R.S. Jakhu, S. Freeland, *The Sources of International Space Law*, Proceedings of the International Institute of Space Law, 2014, 460 ss; *see also* O. Ribbelink, *Article III*, in S. Hobe, B. Schmidtt Tedd, K.U. Schrogl (eds.), *Cologne Commentary on Space Law*, Vol. I, Carl Heymanns Verlag, Cologne, 2009, 64 ss; T. Marauhn, *The Use of Force in Outer Space Articles III and IV of the Outer Space Treaty from the Perspective of General International Law*, in S. Hobe, S. Freeland (eds.), *In Heaven as on Earth? The Interaction*

the United Nations Charter, as well as the maintaining of international peace and security.

Then, Article IV para. 1 of the OST has to be examined as it foresees that only the placement of weapons of mass destruction is prohibited, but that the use of outer space for military purposes, including the placement of conventional weapons in outer space, are allowed.[23] Indeed, the first paragraph forbids to put in orbit around the Earth space objects carrying nuclear weapons[24] or any other kinds of weapons of mass destruction, install such weapons on celestial bodies, or station such weapons in outer space in any other manner. Thus, explicitly referring to nuclear weapons and weapons of mass destruction, the prohibition of Article IV para. 1 applies neither to conventional weapons nor to military or dual-use satellites. In contrast, para. 2 provides that the Moon and other celestial bodies have to be used exclusively for peaceful purposes. The use of the Moon and other celestial bodies shall be exclusively for peaceful purposes specifying that no bases, installations or fortifications can be established on celestial bodies. It provides for the full demilitarization of the Moon and other celestial bodies banning the placement and testing of any type of weapons, the establishment of military installations and facilities, and any kind of military manoeuvres. Nevertheless, the use of military personnel and facilities in the context of scientific and peaceful activities is permitted.

These provisions are broadened by Article 3 of the Moon Agreement,[25] which prohibits any threat or use of force or any hostile act or threat of hostile act on the Moon. It prohibits the positioning of objects equipped with nuclear weapons or mass destruction not only on the Moon, but also in its orbit or trajectory to the Moon.

Thus, Article IV para. 1 provides for a denuclearization regime but not a complete demilitarization of outer space. States remain free to deploy in outer space any type of military satellites and to use outer space for conventional weapons. Moreover, the Article IV does not provide any definition of the terms military uses, nuclear weapons, weapons of mass destruction, or peaceful uses, or made a clear distinction between the terms peaceful and exclusively peaceful.[26]

In this context, some activities carried out by States reflect the limits of Article IV, and the risk of weaponization of outer space. Examples include anti-satellite (ASAT)

---

of Public International Law on the Legal Regulation of Outer Space. 1/2 June 2012, Bonn—Oberkassel, Institute of Air and Space Law of the University of Cologne, 2013, 7 ss.

[23]T. Masson-Zwaan, M. Hofmann, *Introduction to Space Law*, Wolters Kluwer, 2019, 67 ss.

[24]The International Court of Justice in its Advisory Opinion on the Legality of the Threat or Use of Nuclear Weapons of 1996 stated that "nuclear weapons are explosive devices whose energy results from the fusion or fission of the atom".

[25]*Agreement Governing the Activities of States on the Moon and Other Celestial Bodies*, December 1979, 1363 U.N.T.S. 3.

[26]S. Mosteshar, *Space Law and Weapons in Space*, Oxford Research Encyclopedia of Planetary Science, May 2019, 7 ss: https://oxfordre.com/planetaryscience/view/10.1093/acrefore/9780190647926.001.0001/acrefore-9780190647926-e-74?print=pdf.

capabilities,[27] acts of espionage in orbit,[28] jamming signals and cyberattacks. These activities are real, and the ASAT test conducted by China in 2007,[29] the United States in 2008[30] and India in 2019[31] reflect perfectly this trend.

Furthermore, States are developing space defence strategies and policies. In this context, they are establishing "space force" such as in France[32] and in the United-States.[33] Nowadays, States possess significant military or dual-use space technologies. Moreover, a growing number of States and other space actors are expected to develop new capabilities for military reasons.[34]

The current legal regime does not provide a comprehensive system covering all aspects of the military uses of outer space. If it expressly prohibits weapons of mass destruction in outer space, it does not take into account other weapons and "hostile act" that States might engage in outer space in particular due to the dual use nature of space technologies and space missions. Indeed, the neutralization of satellites, their functions or satellite architectures, can be achieved in different ways through different space activities such as on-orbit servicing rendezvous and proximity missions.

1.3 OOS Maneuvers and Risks of Hostile Actions in Space: Issues Tackled by the International Community

Given the limits of Article IV of the OST and the fact that OOS vehicles might represent a risk to the peaceful uses of outer space by causing interferences, jamming or by using a laser against the spacecraft serviced, these elements denote the risk

---

[27] *See* UNIDIR, Space dossier, *Towards ASAT Test Guidelines*, May 2018, 18p: https://www.unidir.org/publication/towards-asat-test-guidelines.

[28] BBC website, *Russia 'tried to spy on France in space'—French minister*, Sept. 2018: https://www.bbc.com/news/world-europe-45448261.

[29] Secure World Foundation, *2007 Chinese Anti-Satellite Test*, Fact Sheet, Nov. 23, 2010: http://swfound.org/media/205391/chinese_asat_fact_sheet_updated_2012.pdf.

[30] https://www.reuters.com/article/us-satellite-intercept-vulnerability/u-s-shot-raises-tensions-and-worries-over-satellites-idUSN2144210520080222; https://www.thespacereview.com/article/3277/.

[31] B. Weeden, V. Samson, *India's ASAT Test is Wake-Up Call for Norms of Behavior in Space*, SpaceNews, April 8, 2019: https://spacenews.com/op-ed-indias-asat-test-is-wake-up-call-for-norms-of-behavior-in-space/; https://thediplomat.com/2019/05/why-indias-asat-test-was-reckless/.

[32] S. Morgan, *Macron announces launch of French Space Force*, Euractiv, July 15, 2019: https://www.euractiv.com/section/aerospace-and-defence/news/mon-ready-macron-announces-launch-of-french-space-force/; Permanent representation of France to the Conference on Disarmament, *Florence Parly unveils the French space defence strategy*, July 25, 2019: https://cd-geneve.delegfrance.org/Florence-Parly-unveils-the-French-space-defence-strategy.

[33] US Space Force Official website: https://www.spaceforce.mil/; *see also* Space Policy Directive-4, *Establishment of the United States Space Force*, Feb. 19, 2019: https://www.whitehouse.gov/presidential-actions/text-space-policy-directive-4-establishment-united-states-space-force/.

[34] F. Lyall, P.B. Larsen, *Space Law A Treatise*, Routledge, 2018, 465 ss.

of weaponization[35] of outer space because OOS vehicle might be considered as a weapon.

In fact, even if OOS capabilities are not weapons at first sight, they can be perceived as dormant threats due to their dual-use nature. Indeed, one can argue that a civil operation may withhold a hostile behavior. Even if OSS programs are not necessarily military space activities, their primary task such as *rendezvous* with and manipulate other space objects afford for potential military maneuver. In this context, OOS operations highlight issues concerning the "weaponization" of space and especially the conditions in which this technology might be used in a non-peaceful way. OOS capabilities might be viewed as an "aggressive" behavior if it used to interfere with another space asset.[36] So, the crucial question is to figure out whether future OOS missions will infringe the principle of peaceful uses of outer space in view of the OOS vehicle's aim.

In this context, OOS missions are discussed within the United Nations Institute for Disarmament Research (UNIDIR)[37] and in the framework of PAROS—*Prevention of an Arms Race in Outer Space*.[38] PAROS has been developed under the auspices of the United Nations General Assembly which passes an annual resolution focused on this issue,[39] and the UN Conference on Disarmament (CD).[40]

Given the fact that OOS vehicles may perform a variety of different tasks, such as human spaceflight docking, orbital assembly, satellite servicing, repair and refueling, inspection, intelligence collection and co-orbital anti-satellite, one can argue that they are not just "dual-use" but "multi-use" that is technology that can serve defensive

---

[35]The term 'weaponization of space' refers to the deployment of weapons in outer space to attack, destroy, or damage objects in outer space, as well as human beings and objects on the Earth. *See also* S. Freeland, *International Law and the Exploration and Use of Outer Space*, in M. Ambrus, R. Rayfuse (eds) *Risk and the Regulation of Uncertainty in International Law*, Oxford University Press, 2017, 77–96.

[36]M. Bourbonniere, *National-Security Law in Outer Space: The Interface of Exploration and Security*, 70 J. Air L.&Com. 3 (2005).

[37]D. Porras, *Shared Risks: An Examination of Universal Space Security Challenges*, Briefing Paper, UNIDIR, 2019, 6 ss.

[38]UNGA Res. 36/97, 9 December 1981; *see also* B.G. Chow, *Space Arms Control: A Hybrid Approach*, Strategic Studies Quarterly, 12(2), 2018, 107–132; C.A. Ford, *Whither Arms Control in Outer Space? Space Threats, Space Hypocrisy, and the Hope of Space Norms*, April 6, 2020: https://www.state.gov/whither-arms-control-in-outer-space-space-threats-space-hypocrisy-and-the-hope-of-space-norms/.

[39]UN Res. 74/32, *Prevention of an arms race in outer space*, UN Doc. A/RES/74/32 (12 December 2019); UN Res. 74/33, *No first placement of weapons in outer space*, UN Doc. A/RES/74/33 (12 December 2019); UN Res. 74/34, *Further practical measures for the prevention of an arms race in outer space*, UN Doc. A/RES/74/34 (12 December 2019).

[40]T. Masson-Zwaan, M. Hofmann, *Introduction to Space Law*, Wolters Kluwer, 2019, 69 ss; G. Alves, *Prevention of an Arms Race in Outer Space: A Guide to the Discussions in the Conference on Disarmament*, United Nations Publications, New York, 1991, 8 ss.

or offensive goals.[41] In other words, the technology can be originally used for civil purposes, and in the same time, be weaponized.[42] Moreover, it is a proliferating space technology[43] used by a wide variety of actors.

One of the main challenges for space security rests on the system of verification of space technology and activities in outer space,[44] in particular due to their dual-use/multi-use nature and the lack of confidence between space actors.[45] This issue has been addressed in the framework of PAROS,[46] especially on the fact that such a system should be able to detect if there is a 'militarily meaningful' violation.[47] According to a UNIDIR Study, "*verifying the on-orbit actions of a space object is easier that verifying its functions*".[48] This is particularly relevant in the case of OOS missions. Over the last years, works and discussions have underscored the necessity to implement significant best practices and standards for OOS in order to avoid misunderstanding and misperception in achieving the mission.[49]

Hence, TCBMs instruments relating to the "dual-use/multi-use" of novel space missions, such as OOS, are crucial to strengthen international cooperation, to build confidence among space actors and to promote the peaceful use of outer space.

1.4 *Transparency and Confidence Building Measures* for On-Orbit Servicing Missions

In this context of potential instability, States will have to clearly prove their intent when engaging in these activities irrespective of whether the mission's aim is commercial or civil.

Given the fact that OOS systems make difficult for States to identify civilian, commercial or military missions, as well as peaceful or hostile acts, it is of utmost

---

[41] D. Porras, *Shared Risks…op.cit.*, 10.

[42] P.J. Blount, *On-Orbit Servicing and Active Debris Removal: Legal Aspects*, in A. Peculjic, M. Tugnoli (eds.), *Promoting Productive Cooperation Between Space Lawyers and Engineers*, IGI Global, 2019, 179–192.

[43] B. Weeden, *The Evolution of Space Rendezvous and Proximity Operations and Implications for Space Security*, United Nations Disarmament Conference, New York, April 12, 2019: https://unidir.org/sites/default/files/2019-12/Brian%20WEEDEN%20-%20UNDC_RPO_Apr2019.pdf.

[44] UNIDIR Space Security Conference 2019, *Supporting Diplomacy: Clearing the Path for Dialogue*, 28–29 May 2019, 4 ss.

[45] D. Porras, *Eyes on the Sky—Rethinking Verification in Space*, UNIDIR, Space Dossier 4, Oct. 2019, 7 ss.

[46] Report of the Conference on Disarmament Subsidiary Body 3: Prevention of an Arms Race in Outer Space, CD/WP.611, 3 September 2018, 2.

[47] D. Porras, *Eyes on the Sky… op.cit.*, 12.

[48] B. Basely-Walker, B. Weeden, *Verification in Space: Theories, Realities and Possibilities*, UNIDIR, Disarmament Forum, vol. 3, 2010, 43.

[49] B. Silverstein, D. Porras, J. Borrie, *Alternative Approaches and Indicators for the Prevention of an Arms Race in Outer Space*, UNIDIR, Space Dossier 5, May 2020, 26.

importance to develop *Transparency and Confidence Building Measures* (TCBMs) in order to ensure that these activities will be carried out in a proper way that is by maintaining safety and strategic stability in outer space.[50]

TCBMs represent significant tools for public and private actors in order to conduct activities in outer space activities.[51] Indeed, the purpose is to clarify the actors' intentions, to enhance transparency through information-sharing, and to implement norms of behavior in outer space.[52]

In particular, many space technologies are of dual-use nature, and consequently a lack of clarity of intent regarding future space missions can result in suspicion and misunderstanding in space activities. The dual-use nature of OOS vehicle might undermine political and strategic stability in outer space. Indeed, the possibility of unsupervised access to space assets by service spacecraft presents security risks. Hence, the main concern is to figure out the vehicle mission's objective.

Future TCBMs relating to OOS missions will have to encompass a number of challenges, including:[53]

(i) Identifying best practices and standards in order to reduce misperception in the mission's achievement;
(ii) Setting up normative behavior allowing to differentiate between civil, commercial RPO for peaceful purposes and potentially hostile RPO;
(iii) Enhancing space situational awareness (SSA) for monitoring and verification.

These may have the aim of implementing norms of behavior for future OOS and RPO space activities while also ensuring other States of the legality of space operations.

In this regard, it must be emphasized the work achieved by the *Consortium for Execution of Rendezvous and Servicing Operations* (CONFERS) which the aim is to elaborate and to implement non-binding technical and legal standards for on-orbit servicing[54] such as:

---

[50]Rep. of the Group of Governmental Experts on Transparency and Confidence-Building Measures in Outer Space Activities, U.N. Doc. A/68/189 (2013).

[51]UN Res. 74/67, *Transparency and confidence-building measures in outer space activities*, UN Doc.A/RES/74/67 (12 December 2019).

[52]P. Martinez, R. Crowther, S. Marchisio, G. Brachet, *Criteria for developing and testing Transparency and Confidence-Building Measures (TCBMs) for outer space activities*, Space Policy (2014), 1–7; see also J. Robinson, *Space Transparency and Confidence-Building Measures*, in KU Schrogl, P. Hays, J. Robinson, D. Moura, C. Giannopapa (eds), *Handbook of Space Security*, Springer, 2015, 291–297.

[53]B. Weeden, *The Evolution of Space Rendezvous …op.cit.*, United Nations Disarmament Conference, New York, April 12, 2019.

[54]CONFERS: https://www.satelliteconfers.org/about-us/; see also A.S. Martin, S. Freeland, *Food for Thought: Relevant Legal Standards for On-Orbit Servicing Missions and Rendezvous and Proximity Operations*, ZLW 69, 2(2020), 308–327.

(i) Guiding principles for commercial rendezvous and proximity operations and on-orbit servicing[55];
(ii) Recommendation for design and operational practices[56]; and
(iii) Document on on-orbit servicing mission phases[57] which establishes a baseline of mission phases describing the functions of all missions.

1.5 Closing Remarks—Towards OOS/RPO "Multi-use" Capabilities

The issue of the dual-use nature of OOS missions is of particular importance in order to ensure the peaceful uses of outer space and the long-term sustainability of space activities. This technology can not per se be considered as a weapon. Nevertheless, even though the legal regime does not prohibit OOS technology, it prohibits certain uses that may interfere and harm with the activities of other States.

These technologies have the capacities to be very destabilizing due to their potential "multi-use" and the variety of actors that can be engaged in these activities. As a result, States, agencies and private companies have to be clear regarding their specific intentions when undertaking OOS/RPO activities irrespective of whether the mission's aim is military, civil or commercial. The lawfulness of these space operations must be guaranteed in order to avoid tensions and conflicts.

The way towards the adoption of TCBMs seems the most suitable because if these instruments are legally non-binding, they are from a political point of view. Moreover, States can decide to implement such voluntary frameworks as domestic rules and thus could become a consistent practice. Industries can as well adopt some standards and guidelines to commit on-orbit servicing missions as noted above. Furthermore, it is necessary to reduce the potential of mistrust and misunderstanding in the achievement of "dual-use/multi-use" missions in outer space especially on-orbit servicing.

---

[55]CONFERS Guiding Principles (updated Nov. 2018): https://www.satelliteconfers.org/wp-content/uploads/2018/11/CONFERS-Guiding-Principles_7Nov18.pdf.

[56]CONFERS Recommended Design and Operational Practices (updated Oct. 2019): https://www.satelliteconfers.org/wp-content/uploads/2019/10/CONFERS_Operating_Practices.pdf.

[57]CONFERS On-Orbit Servicing Mission Phases (updated Oct. 2019): https://www.satelliteconfers.org/wp-content/uploads/2019/10/OOS_Mission_Phases.pdf.

Anne-Sophie Martin is a Doctor of Law specialized in International Law and Space Law. Her doctoral research focused on the legal aspects of dual-use satellites. She received her LL.M in Space Law and Telecommunications Law from the University of Paris-Sud XI (France) and her PhD from Sapienza University of Rome (Italy). On August 2017, she attended the Centre for Studies and Research of The Hague Academy of International Law. Since March 2019, she is, as volunteer, visiting researcher within the 'Centre for a Spacefaring Civilization'. Since July 2019, she is a Legal Council's member of the organization 'For All Moonkind'.

Chapter 2
Risk Management and Insurance of On-Orbit Servicing

Katarzyna Malinowska

Abstract The concept of on-orbit servicing has been present in the space industry for a long time. Although the initial attempts were quite successful, they took place 20 years ago and in terms of commercial activity were followed by a long period of stagnation. Though, during that time technology has substantially developed, especially in terms of robotics, with subsequent tests demonstrating spectacular successes and growing technological readiness. Other circumstances have apparently changed as well, among them the massive production of space debris, rapid shrinking of the orbital slots as well as changes in the space business models requiring an ever more agile approach. All of these are not, however, solely within the circle of interest of the satellite operators. Just as involved are the regulators and the policy makers, and OOS becomes one of the tools for achieving the objectives of sustainable space development promoted at a global level. All of this also concerns the insurance industry, and it should come as no surprise that there is no space mission without financing, and no financing without insurance. It is well known that insurers have accompanied space ventures since the very beginning. Their special role is, however, not limited to supporting financial schemes. There can be no doubt that the insurance industry has a vital role to play in the risk management processes, as it initially developed the risk management concepts and tools that were subsequently applied in all the industries. This is also the case with on-orbit servicing. In effect, a very interesting feedback loop may appear, whereby on-orbit services and the insurance market give each other a synergic burst of new possibilities. The author intends to explore the way in which the new type of space activities and insurance interact, while also enhancing sustainable development at the same time. Nowadays, OOS seems to be an emerging concept also commercially, welcomed with much hope and enthusiasm from the side of the space industry as well as space-faring countries. The potential of this idea is huge, as it can resolve at least some of the many sensitive issues, space debris being one of the most important. No doubt it first of all requires reliable technology, but also the management and legal aspects should be treated with lots of attention. Bearing that in mind, this chapter will focus on the risk management and insurance aspects.

K. Malinowska (✉)
Kozminski University, Warsaw, Poland
e-mail: katarzynamalinowska@kozminski.edu.pl

© The Editor(s) (if applicable) and The Author(s), under exclusive license to Springer Nature Switzerland AG 2020

A. Froehlich (ed.), *On-Orbit Servicing: Next Generation of Space Activities*, Studies in Space Policy 26, https://doi.org/10.1007/978-3-030-51559-1_2

Thus the chapter is not aimed at discussing the technical possibilities of on-orbit servicing, nor at providing a detailed risk assessment, but instead will focus on risk management, with special attention given to insurance.

2.1 Introduction

The concept of on-orbit servicing (hereinafter also referred to as "OOS") has been present in the space industry for a long time. Although the initial attempts were quite successful, they took place 20 years ago and in terms of commercial activity were followed by a long period of stagnation.[1] Though, during that time technology has substantially developed, especially in terms of robotics, with subsequent tests demonstrating spectacular successes and growing technological readiness, applied, for example, for the needs of the ISS. Other circumstances have apparently changed as well, among them the massive production of space debris, rapid shrinking of the orbital slots[2] as well as changes in the space business models requiring an ever more agile approach. All of these are not, however, solely within the circle of interest of the satellite operators. Just as involved are the regulators and the policy makers, and OOS becomes one of the tools for achieving the objectives of sustainable space development promoted at a global level.

All of this also concerns the insurance industry, and it should come as no surprise that there is no space mission without financing, and no financing without insurance. It is well known that insurers have accompanied space ventures since the very beginning, the first insurance coverage being written in 1965 (though only for the pre-launch stage). Their special role is, however, not limited to supporting financial schemes. There can be no doubt that the insurance industry has a vital role to play in the risk management processes, as it initially developed the risk management concepts and tools that were subsequently applied in all the industries. This is also the case with on-orbit servicing. On the one hand, on-orbit services will need to be protected with respect to both property damage as well as third party liability. This conviction results from both the commercial needs of the financing schemes, as well as from the regulatory (national) measures imposing compulsory third party liability insurance ("TPL") insurance as one of the risk management measures. On the other hand, on-orbit servicing creates not only a challenge for insurers to cover new ultra-hazardous activities, but also has the potential to become a new driver on the volatile

[1]These concerns the successful Skylab Mission, Telescope Hubble as well as the recovery of the Palapa B2 and Westar 6 satellites. Frontiers of space risks. Natural Cosmic Hazards and Societal Challenges, ed. R.J. Wilman, Ch. J. Newman. Francis & Taylor Group 2018, p. 179. On-Orbit Satellite Servicing Study. Project Report, NASA Goddard Space Flight Center, October 2010, p. 16.

[2]M. J. Losekamm, et al., Legal and Political Implications of Future On-Orbit Servicing Missions, 66th International Astronautical Congress, Jerusalem, Israel. Copyright c 2015 by the Space Generation Advisory Council; R. Parker, On-orbit servicing—an insurer's perspective, Room 2/2015 (release 27.02.2015).

space insurance market. It can affect the scope of the coverage and can change the paradigm from all risk to named perils insurance, allow for a new concept of calculating the loss formula, including the salvage of the satellite clauses and numerous other concepts that may bring space insurance closer to the 'common' insurance business, more affordable to more insurers and, as a result, to more satellite operators. In effect, a very interesting feedback loop may appear, whereby on-orbit services and the insurance market give each other a synergic burst of new possibilities. The author intends to explore the way in which the new type of space activities and insurance interact, while also enhancing sustainable development at the same time.

Nowadays, OOS seems to be an emerging concept also commercially, welcomed with much hope and enthusiasm from the side of the space industry as well as space-faring countries. The potential of this idea is huge, as it can resolve at least some of the many sensitive issues, space debris being one of the most important. No doubt it first of all requires reliable technology, but also the management and legal aspects should be treated with lots of attention. Bearing that in mind, this chapter will focus on the risk management and insurance aspects. Thus the chapter is not aimed at discussing the technical possibilities of on-orbit servicing, nor at providing a detailed risk assessment, but instead will focus on risk management, with special attention given to insurance. The scope of the chapter encompasses the identification of the risks related to on-orbit servicing in the context of the possibility of implementing business/management mitigation techniques, as well as an analysis of the insurance aspects of the risk management thereof. It will also address the implication of on-orbit servicing for the insurance industry and the insurability of the space risks.

2.2 Risks Involved with On-Orbit Servicing

2.2.1 Technological Context and Risks of OOS

There is no established terminology that would constitute an uniform background for discussing on-orbit servicing. It has been described as "*on-orbit activities conducted by a space vehicle that performs an up-close inspection of, or results in intentional and beneficial changes to, another resident space object*",[3] or as "*the on-orbit alteration of a satellite or its orbit after its initial launch, using another spacecraft to conduct these alterations. Examples include relocating the satellite to a new orbit, refuelling, repairing broken parts, replacing parts, deploying systems that failed to*

[3]J.P. Davis, J.P. Mayberry, J.P. Penn, On-orbit servicing: inspection, repair, refuel, upgrade, and assembly of satellites in space.

deploy after launch, and cleaning components."[4] Analysing OOS step by step in terms of the mission phases, it includes launching a servicing spacecraft, orbiting (orbital manoeuvring), getting in touch (or at least in the near proximity) with the target satellite, autonomous rendezvous and docking and then possibly de-orbiting. As regards the service actions, they may include robotic manipulation, modification, refuelling, commodities replenishment, repair, upgrade or correction of mechanical failures and assembly. Depending on the method applied, the target satellite may be cooperative or uncooperative, depending on the technical features thereof, such as the possibility of docking and transferring information.[5] As regards the benefits of on-orbit servicing, the range of the possibilities include non-contact support (consisting mainly of inspection or wireless support), orbit modification and maintenance (such as repositioning, etc.), creating new space objects, prolonging the life of the existing satellite and restoring functionality and the ability to achieve profits, as well as debris mitigation.[6] It is clear from this that OOS includes a full range of already known space activities, as well as those that are completely new, and which will have to be taken into account in the risk management measures. In this respect it is however also stressed, that on-orbit servicing missions are still lacking of achieving critical mass to make realistic assessments.[7] Although technology is key, which should be dealt with by the risk management tools, it should also actively respond to the risk assessment results, which should be performed on a continuous basis.[8]

The first issue to clarify when discussing risk management is the meaning of risk. It seems necessary to distinguish risk from the notion of peril or hazard. The latter mean the cause of a potential loss, and factors contributing to the occurrence of a loss, or increasing the severity of the loss or conditions affecting the perils. As regards the definition of risk for purposes of this chapter, we can use many examples, such as what is included in the US Code of Federal Regulations ('CFR'), which describes the risk as 'a measure that accounts for both the probability of occurrence of a hazardous

---

[4]S. A. Carioscia, B. A. Corbin, B. Lal, Roundtable Proceedings: Ways Forward for On-Orbit Servicing, Assembly, and Manufacturing (OSAM) of Spacecraft, 2018 Institute for Defense Analyses.

[5]The best example of a cooperative space object is the ISS; See also, for example, the project MEV (Vivisat by Orbital ATK and US Space; D. Benoussan, TeSeR—Technology for Sel-Removal of Spacecraft (Project under Horizon 2020 under grant agreement No. 687295), p. 13); J.P. Davis, J.P. Mayberry, J.P. Penn, On-orbit servicing: inspection, repair, refuel, upgrade, and assembly of satellites in space; On-Orbit Satellite Servicing Study. Project Report, NASA Goddard Space Flight Center, October 2010, p. 11.

[6]Ibid, p. 3; see also R. Reesman, Assurance through insurance and on-orbit servicing, The Aerospace Corporation 2018.

[7]P. Colmenarejo, M. Graziano, Towards Cost-Effective On-Orbit Servicing/ADR Using Modular and Standardized Approach, IAC-19-A6.10/B4.10 x 53000.

[8]W. Connley, Integrated Risk Management Approach within NASA Programs/Projects, https://ntrs.nasa.gov/search.jsp; J.S. Perera, L.B. Johnson, The Risk Management for the International Space Station, Proceedings of Joint ESA-NASA Space Flight Safety Conference, ESTEC, Noordwijk 2002.

event and the consequence of that event to persons or property.'[9] Identifying the hazards, hazardous events risk factors and the subject matter of the risk is the first step in effective risk management and insurance.

For the purposes of risk management and insurance analytics, the risks may be divided into categories related to the object suffering the damage, i.e. by property and persons, mirrored accordingly in the category of property risks and personal risks. These include the loss, damage or destruction of property related to a space project (i.e. satellite, launch vehicle, ground facilities), as well as the property of third parties, such as ships and aircraft (in the event of a collision during the launch), or any other property in the event of a space object hitting the ground, increased by the consequential loss of profits and pure financial loss.[10] Damage to persons includes manned space flights, persons involved in some other way with the space activities (e.g. launch facility staff), as well as innocent by-standers.[11]

Turning briefly to the features of space risks and the underlying hazards, they are associated primarily with "anything outside of Earth's atmosphere that can cause harm to people or property."[12] It is obvious that the risk related to space operations is not limited to cosmic threats, but also includes those that can occur on Earth. This is supported by the broad approach taken by the regulation of space activities included in the legal provisions and contractual practice.[13] Considering the range of threats posing a potential danger, OOS would rather remain in the same category as ultra-hazardous activity. These are a mixture of technological, human and natural perils, categorised differently. The most obvious are threats due to the technology that must be used in order to carry out a space mission, relating to the propellants, transforming into kinetic energy, reaching a high velocity and encountering enormous

[9]See also E. Baranoff, who gives a similar definition relating risk to the consequences of uncertainty, as well as E. Vaughn: E. Vaughn, Fundamentals of Risk and Insurance, John Wiley & Sons, Inc., 2008, p. 5; E. Baranoff, Enterprise and Individual Risk Management, Creative Commons 2012, p. 22 as well as D.M. Gerstein et al., Developing a Risk Assessment Methodology for the National Aeronautics and Space Administration, 2016, Library of Congress, p. 7.

[10]Blassel P., *Space projects and the coverage of associated risks*, The Geneva Papers on Risk and Insurance, Vol. 10, No. 35, 1985, p. 72—proposes another division, distinguishing separately: loss of property, damage to property, a failure to achieve the proper orbit, a partial or total failure of the satellite or payload and a loss of revenues. However, according to the author, the above division includes damage as well as the risk from the occurrence of which the damage results.

[11]In total, over 200 people have been killed by rocket explosions. Apart from the Challenger space shuttle, the majority of the accidents causing death occurred on the ground during the ground processing of the launch operation, or during re-entry; S.R. Jakhu, T. Sgobba, P.S. Dempsey (2011), The need for an integrated Regulatory Regime for Aviation and Space, p. 13.

[12]Ross S, *Risk Management and Insurance industry perspective on cosmic hazards*, in: Handbook of cosmic hazards and planetary defence, 2015, p. 2.

[13]For example, the launch activities as proposed by the UN "shall be defined as those activities undertaken to place or try to place a launch vehicle and any payload in a suborbital trajectory, in Earth orbit in outer space, or otherwise in outer space.".

friction while traversing the atmosphere.[14] Any accidents related to the release of the propellant during the launch stage are known to cause explosions, debris, fire, and toxic vapour clouds, though the severity of the above depends also on the type of propellant. Depending on the circumstances of the case, some of the above hazards may appear competitively or may preclude each other, depending on the vehicle design, accident location, the vicinity of people and property, a failure mode, type and amount of propellant, as well as environmental conditions. The probability of any of these hazards appearing is dynamic and changes during the space mission phase.

The satellite operation stage is subject to many risks that do not end with a successful launch, but only really begin then, lasting until the satellite is functional, and even thereafter, where the obligation to de-orbit and the risk of re-entry comes into question. It is under constant exposure to environmental hazards. It is common knowledge that operations in outer space are performed in extreme environmental conditions, which cannot be avoided, but can only be mitigated to a certain extent. The environment of outer space is a hostile one to satellites: with extremes temperatures, cosmic radiation and electromagnetic fields, a vacuum, etc. all causing a significant challenge to the lifetime of a satellite and its functionality (including of an unintended change in orbit).[15] This cannot be changed in spite of the advancement of technology, though no doubt the possibilities of effective protection are increasing and satellites are already designed to have high resilience to the hazards of space. The perils most often resulting in damage suffered by a satellite are listed by Lloyd's of London in the RDS (Realistic Disaster Scenarios), which include four potential risks, i.e.: an anomalously large solar energetic particle event affecting many satellites, a generic defect causing undue space weather sensitivity in a class or classes of satellites, a generic defect causing unforeseen failures in a class or classes of satellites, and finally collision with debris.[16] The fact it cannot be avoided is, by the way,

---

[14]Propellants are also used during the satellite operations stage, but in significantly lower quantities, which also lowers the probability of technology-related failures. That is also why the risk related with the use of propellant decreases along with the flight time and the consumption of the propellants; A. Soucek, *International Law* in: Outer Space in Society, Politics and Law, A. Soucek, Ch. Brunner, Springer Wien New York 2011, p. 338.

[15]In this respect, there are at least three types of radiation that are taken into account and which may vary, depending on the orbit. These are van Allen belts—captured by the Earth's magnetic field, particles sent by the sun during solar storms and galactic cosmic rays; Kleiman J., Lamie J. K., Carminati M.-V., *The law of spaceflight, A guidebook for new space lawyers, 2012*, p. 20; Pelton J., *Satellite communications*, Springer Science & Business Media, Arlington 2011, p. 30.; also A. Soucek, International Law in: Outer Space in Society, Politics and Law, A. Soucek, Ch. Brunner, 2011, p. 337.

[16]M. Williamson, Commercial Space Risks, Spacecraft Insurance and the Fragile Frontier, in: Frontiers of space risks. Natural Cosmic Hazards and Societal Challenges, ed. R.J. Wilman, Ch. J. Newman. Francis & Taylor Group 2018, p. 149.

one of the most important reasons for implementing OOS (degradation of the solar panels, etc.).[17]

An example of a hazard of growing importance is debris, which can affect a space mission at any stage. These may be space debris located in the Earth's orbit, or even debris created at the launch phase as a result of jettisoning subsequent stages of a launch vehicle. It is estimated that there are more than 22,000 objects, qualified as debris, being tracked,[18] and a collision with even a tiny piece of debris may damage, if not destroy the satellite, resulting in even more debris.[19] This type of hazard produces a risk to the assets, including spacecraft costs, launch vehicle costs, insurance and own capital costs, as well as third party liability, pure financial loss such as manufacturers' incentives, contract obligations and business interruption risks (operating and extra expenses, loss of revenues, etc.) Risk management tools applied so far in that field consisted mainly of limiting the volume of the produced debris. Active debris removal concepts interact with the possibilities offered by OOS.[20] Apart from that, new types of hazards should be taken into account and distinguished, especially in the satellite operation stage. These mostly concern intentional interference and cyberattacks. There can be no doubt that the space industry is also exposed to legal, political and commercial, operational and other risks related to any business activity. Due to the transboundary character of space activities, and the high involvement of the states, these risks seem to be more internationally correlated than in other types of business activity.

It should be underlined that these threats are not characteristic to OOS specifically, but to all types of space missions. This said, it is likely that OOS will be exposed to similar risks as 'conventional' space activities, adding 'only' the new types of spacecraft that may suffer the damage. Due to the robotic nature of OOS activity, we can assume that human risks will not play a substantial role in the overall risk assessment (as was the case in the early OOS missions which were performed with the help of

---

[17] M. Hapgood, Space weather, in: Frontiers of space risks. Natural Cosmic Hazards and Societal Challenges, ed. R.J. Wilman, Ch. J. Newman. Francis & Taylor Group 2018, pp. 38–49.

[18] ESA 2015, CLEANSAT: New satellite technologies for cleaner low orbits.

[19] This is known as the Kessler syndrome (space-asset destructive chain reaction) following the name of a NASA expert Donald Kessler, who in 1978 first discussed the potential of orbital debris becoming self-perpetuating. It was concluded that collisions of satellites and spent rocket bodies would eventually form the dominant source of orbital debris in LEO. It was predicted that debris from collisions would collide with other satellites and rocket bodies and create even more debris. As a result of this chain reaction, the risk to satellites in certain regions of space would increase exponentially over time, even without further launches into those regions. In a 1991 paper, Kessler used the term "collisional cascading" to describe this process. This has created the widely used term "Kessler syndrome"; see for example EU (2013) MEMO/13/149.

[20] C. Colombo, F. Letizia, M. Trisolini, H. Lewis, Space Debris, in: Frontiers of space risks. Natural Cosmic Hazards and Societal Challenges, ed. R.J. Wilman, Ch. J. Newman. Francis & Taylor Group 2018, pp. 112–120.

space shuttles and astronauts).[21] As regards the risk implications for the servicing spacecraft, the inherent nature of in-orbit risks could possibly be distinguished from those coming from OOS operations and the features of the target satellite. In this respect, the question arises whether the risks related to each of them are of such a substantially different type, if we consider the property damage, loss of revenues or liability, that it will change any of the risk management paradigms? This question should be answered by the engineers, so that risk managers may follow by adopting adequate measures.

Analysing the risks involved with OOS in more detail, two aspects should be distinguished. The risk and management thereof from the point of view of the satellite being serviced, and, from the other side, the risks related to the on-orbit servicing mission. While it seems that the majority of the risks in both aspects are similar and well recognised, there are certainly also those that do not appear on a daily basis with no OOS applied. Therefore, in order to analyse the issue in an analytical way, the risks related to conventional in-orbit satellite operation and the risks related to on-orbit servicing should be recognised separately. This seems important also in the context of the insurance coverage, which tackles the target satellite separately, the OOS spacecraft separately and potentially the OOS mission separately, with respect to the liability risks. This division reflects the types of coverage offered by insurers. An analysis of the risks related to the conventional operation of a satellite is also one of the best forms of evidence demonstrating that OOS will be indispensable in the future of the space industry.

It is worth noting that each of the space actors face slightly different risks in relation to satellite operations in-orbit, meaning that each perceive OOS from slightly different perspectives. The satellite operator's exposure concerns assets, revenues, expenses and liabilities, all those risks being concentrated during the in-orbit phase. Satellite manufacturers face the risk during the manufacturing process until the legal transfer of the satellite to the operator, facing the risk to assets, liability, expenses, and financial incentives—depending on the contract with the client. Launch service providers are exposed to the liability and obligation to relaunch (under a relaunch guarantee). Even this rough outline shows that, despite having different roles and being directly exposed at different stages of the space project, the consequences of a misperformance at any of the pre-orbital stages cumulate at the in-orbit stage involving all of them. As a result, each of them might be interested in applying OOS, even if only to protect their very particular business interests. Apart from that, the general interest concerns all of them, though the main pressure certainly comes from the policymakers, such as UN, space agencies, as well as academic circles, etc. It is related not only with the common belief in the 'heritage' represented by outer space, but also with the liability that is ultimately borne by the launching states.

[21] On-Orbit Satellite Servicing Study. Project Report, NASA Goddard Space Flight Center, October 2010, p. 25.

2.2.2 Legal Context and Risks of OOS

Although this chapter does not focus on the legal aspects of OOS, that problem cannot be totally avoided. The legal context is important for the risk management and insurance for several reasons. Though the risk and its management lie within the domain of economical and management science, there is also no doubt that the law affects the issues of risk itself.[22] This can happen firstly by imposing an obligation to apply the technical measures to avoid or mitigate the risk as a prerequisite of licensing—which per se is perceived as a risk factor,[23] and secondly, by allocating the risk to specific entities. Thus the (space) law can implicate a legal obligation to compensate damage suffered by a related party (known as a second party) or a third party[24] in those situations where technical measures were not sufficient to avoid the damage, or were not applied properly. It may also exclude or limit the liability. The way such an allocation is made, on the basis of statutory (for example in France and the US) or contractual provisions, is an obvious risk management tool of a legal nature. As such, it has an impact on insurance, i.e. the insurable interest and the type of insurance coverage applicable.[25]

As regards OOS, no doubt both issues are to be resolved, i.e. the technical requirements being part of the licensing process,[26] as well as the liability regime. The question that should be resolved is whether the measures applied at the moment are

---

[22]Risk management for the purposes of space projects is used in the meaning adopted by IAASS: "*Risk management* is a systematic and logical process to identify hazards and control the risks they pose." One of the important reasons for adopting space law in general, is the authorisation of space activities, during which the states have a chance to verify the technical tools adopted by space entrepreneurs, and during the continuing supervision of space activities to check the application of the measures accepted at the authorisation stage.

[23]M. Laisne, Space Entrepreneurs: Business Strategy, Risk, Law and Policy in the Final Frontier, 46 J. Marshall Law Review 1039 (2013), Issue 4. H. Brettle, J. Forshaw, J. Auburn, C. Blackerby, N. Okada, Towards future Debris Removal Service: Evolution of an ADR Business Model, IAC-19, A6, 10-B4.

[24]The concept of second and third party risk reflects the potential liability to related (second) parties or (unrelated) third parties. The circle of 'related parties' is defined in a broad way and it includes the whole chain of contractors and subcontractors, where the main criterion is the involvement of an entity in the same space project. From this point of view, also the states are usually included in the notion of the second party, due to the ownership of the space port facilities, ownership of the space object to be launched into outer space, or simply by the international responsibility imposed by OST.

[25]The system of allocating risk in space activities requires applying the criteria of the entities involved in the launch operations, according to which there are first party, second party and third party risks. See also Mendes de Leon P., van Traa H., *The practice of shared responsibility and Liability in Space Law*, Amsterdam Center for International Law, Shares Research Paper 70, 2015, pp. 19–23, available at www.sharesproject.nl, accessed 18 August 2016.

[26]Looking only at the UKSA requirements on risk assessment of quantitative and qualitative nature, where the technical and risk mitigation requirements are explained in detail: https://www.gov.uk/guidance/apply-for-a-license-under-the-outer-space-act-1986, accessed on 7 March 2020; see also T. Harris, K. Memon, G. Glasgow, In-orbit risk assessment in the era of New Space, First International Orbital Debris Conference 2019.

sufficient to encounter OOS activity in both aspects, or whether new concepts must be implemented. The subject certainly requires in-depth analysis and has been included in some academic research projects.[27] It should be assessed on a horizontal basis including the spectrum issues (as OOS will possibly not need an orbital slot), export control requirements and impediments (restricted transfer of data), the protection of intellectual property rights, and some others. With respect to the liability regime, the concepts of changing the fault-based regime into a risk-based regime during the in-orbit phase are multiplying.[28] So far, under the existing regime it seems that OOS is just a type of space activity, as defined by the majority of the national space laws, so liability for damage is essentially fault-based, due to the core OOS operation being performed in-orbit. That is a result of the definitions of space activities included in the specific regime of the space law and authorisation requirements,[29] which focus on the 'upstream' sector, to which OOS inevitably belongs. Doubts are raised, however, whether the existing regime fits the features of OOS, both at a licensing level and with respect to the liability regime. The manner of structuring both of these aspects is part of the risk management process and will have an impact on the insurance of the risks related to the OOS in terms of the risk exposure for insurers. This may include, for example, the compulsory nature of TPL insurance for the in-orbit stage, nowadays not even a rule. The legal aspects of OOS in terms of the insurance industry entail also questions about the licensing of insurers, where they are to be involved in OOS. Is the existing regime sufficient, or should there be some 'legal incentives'?

As we know, the legal regime has been primarily structured in the Outer Space Treaties, providing for the international responsibility of the States Parties to authorise and supervise space activities, and consequently to impose liability for damage caused by space objects with the help of which the space activities are carried out in the launching state. This basic principle of the international liability of the state was subsequently clarified in the Liability Convention. According to Article II of the LC, launching states are internationally liable for damage caused by space objects. If damage is caused by the space object on the surface of the Earth, or to an aircraft in flight, the liability to compensate is absolute (regardless of the proof of fault). If damage is caused to another space object in space (Article III LC), the launching state is only liable if the damage is due to fault on the part of the launching state

[27] M. J. Losekamm, et al., Legal and Political Implications of Future On-Orbit Servicing Missions, 66th International Astronautical Congress, Jerusalem, Israel. Copyright c 2015 by the Space Generation Advisory Council.

[28] M. J. Losekamm, et al., Legal and Political Implications of Future On-Orbit Servicing Missions, 66th International Astronautical Congress, Jerusalem, Israel. Copyright c 2015 by the Space Generation Advisory Council. C. Santos, M. Rhimbassen, On Orbit Servicing as Space Resource. Liability Challenges, http://chaire-sirius.eu/wp-content/uploads/2018/11/ppt-6-nov-OOS-liability.pdf, ast access 28 February 2020

[29] Which in turn are activities that are inherently and strictly related to the environment of outer space in a functional approach, i.e. which are conducted after leaving the ground with the aim of reaching a level of space not reachable by conventional aircraft.

or its nationals. The OST and the LC regulate the management of the liability risk just at an international level. It is for the States Parties to incorporate that regulation, possibly in a more precise way, to their national laws. The role of national laws is essential for the effective protection of public safety, property and environment. It is not only a correlative with fulfilling the international duties of the launching states,[30] but the only enforceable tool to be imposed on space operators. The issue of environmental protection in outer space becomes particularly important in the context of space debris. While it is difficult to implement new binding measures on international level, it is up to the countries to adopt such legislation that would respond to the needs of the sustainable development. It needs shifting the focus from perceiving liability regime solely as a compensatory instrument and to stress its preventive impact.[31] Regulatory risk management powers seem to be obvious in this respect.

An analysis of the allocation of risk in space activities, very helpful for the purposes of managing them, requires categorising the risks in relation to the criteria of the entities involved in the launch operations, according to which there are first party, second party and third party risks. Third party risk can be easily defined from a legal point of view, and has been subject to national and international legislation in the context of ensuring the best possible protection for the victims of space activities. The possibility of contractually shaping the allocation of third party risk is limited and may have effect only between the parties to the contract, while innocent and unrelated victims are protected by the strict provisions of international and national (where enacted) laws. Third party risk has been regulated by the obligatory provisions of law, deriving primarily from the outer space treaties (namely the LC and OST), and implemented in almost all the national acts regulating commercial activity in space. Undoubtedly, the regulation of government control over space activity is inherently related to the need to protect third parties against the risk of damage resulting therefrom, and the need to introduce a mechanism of imposing responsibility and liability over the entities gaining profits from space activity[32] in such a way that the government of the launching state has the possibility of shifting the burden of financial liability for damage caused, which is imposed on the state by the space treaties. *The second party risk* in space projects is perceived in the context of the liability of the participants of the space mission between each other. This type of risk refers to and is mainly regulated by the legal and contractual provisions binding

---

[30] Dempsey P. S., *National Laws Governing Commercial Space Activities: Legislation, Regulation, & Enforcement*, in Northwestern Journal of International Law & Business, vol. 36, 2016, p. 19.

[31] See more on the 'compensatory logic and preventive logic' of the liability for environmental damages—the view expressed in the ICJ Judgement Hungary v Slovakia, *CJ Reports 1997, 78, para 140* in: W. Munders, Active Debris Removal, International Environmental law and the Collective Management of Risk: Foundations of an International System for Space Traffic Management, Working Paper No. 200—April 2018.

[32] Kerrest A., *UN Treaties on outer space, L.C. and licensing regimes*, in: Actions at national level, UN Korea Workshop, pp. 236–249.

upon the parties involved,[33] though it must be stressed that it is understood broadly. In addition to the contracting parties, the second party risk usually also involves the government, which, even where it does not participate directly in the space activity, is also exposed to risk.[34] The identification of second party risks, though to some extent depending on the legal regulations imposed by national laws, will mostly derive from the organisational structure of the operations including ownership issues, and contractual relations between all the parties involved.[35] In turn, *first party risks* are typically those risks that are absorbed by the respective parties to the space operation in such a way that each one assumes the risk of the loss of its own property, and all the consequences resulting therefrom, without the possibility of shifting it to other parties of the space project via contractual liability clauses or claims in tort, thereby substantially limiting the scope of second party risks.

The contractual system of risk allocation consists of inter-party waivers of liability, accompanied by hold harmless clauses, and flow-down provisions, applied jointly. This results mostly from the standard contracts, as international and national legal regulations, with very few exceptions (US and France), do not regulate the risk allocation schemes.[36] Due to the above risk allocation regime, second party risks are substantially mitigated on a contractual or regulatory basis and not insurance. The question of the liability between the contracting parties is one of the crucial aspects for OOS operations. There will emerge a completely new type of contractual relation, where the risk exposure is of a type not met before. Are cross waivers still the best solution? If not, it may result in a demand for a new type of insurance coverage for space activities.

All the above leads to the conclusion that the risk management should be an integrated activity that covers entire project life and takes into account perspectives of all the stakeholders. Thus all of them, including insurers and capital providers should be involved in the process as it enable the project manager to take the more informed decision based on the identification of acceptable and non-acceptable risks and the mitigation possibilities.[37]

---

[33] A similar context of second party risks was presented by Prof T. Tanja Masson-Zwaan—*Liability & insurance in air & space law; Regulation of suborbital flights in Europe*—ICAO/UNOOSA Aerospace Symposium, Montreal Canada, 18–20 March 2015.

[34] Hermida, J., *Commercial Space Launch Services Contracts in France and the United States of America*, Rev. dr. unif. 2004-3, p. 541.

[35] See also Kayser V., *Launching Space Objects: Issues of Liability and Future Prospects*, in Space Regulations Library, vol. 1, Kluwer Academic Publishers, New York, Boston, Dordrecht, London, Moscow 2001, pp. 7–8.

[36] Beer T, *Launch services agreements—anything new, Launch services agreement*, in: Project 2001, Legal Framework for the Commercial Use of Outer Space: Karl-Heinz Böckstiegel (Ed.); Carl Heymans Verlag, Cologne, 2002, p. 129.

[37] W. Connley, Integrated Risk Management Approach within NASA Programs/Projects, https://ntrs.nasa.gov/search.jsp; More on risk-informed decision making—D. M. Gerstein et al., Developing a Risk Assessment Methodology for the National Aeronautics and Space Administration, 2016, Library of Congress, pp. 15 and 58 et subsq.

2.3 Insurance as a Risk Management Tool of On-Orbit Servicing

Insurance is known to be one of the oldest and best developed risk management tools. No doubt, the risk assessment methods developed by insurers as a part of their core underwriting tasks, nowadays serve as a pattern for general risk management in daily business life. Insurance, known since the beginning of early space ventures, includes all the stages of the space missions, starting from the manufacturing process until reaching the in-orbit stage. Space insurers, who are also space engineers, have developed a unique insurance type of coverage and they can be perceived as deterrent to the growth of the industry.[38] Based on the all-risk concept, space coverage is a mixture of various types of insurance.[39] An effective insurance coverage in place is a condition precedent of many financial schemes, as it is perceived as a stabilising factor for risky commercial space activity.[40] Even for governments and for the EU, the external financing and insurance of space projects is becoming an increasingly important tool of risk management, in spite of its costs.[41] The insurance of space risks is very often a part of the financing scheme for the whole space mission.[42]

Space insurance is a highly specialised branch of insurance law and, from a regulatory point of view, it still seems to be a kind of novelty on the insurance market. Taking the EU law approach as an example, there is no regulation explicitly concerning space risks, which for the time being can only be classified as a type of transport insurance.[43] The question is also whether space third party liability

---

[38]This function of insurance is not specific in the space industry but in many others innovative ventures. M. Laisne, Space Entrepreneurs: Business Strategy, Risk, Law and Policy in the Final Frontier, 46 J. Marshall Law Review 1039 (2013), Issue 4.

[39]K. Malinowska, Space Insurance. International Legal Aspects, Kluwer 2017, p. 287. AON, Space Q1 2020 market report.

[40]Meredith P., *Space insurance Law—with a Special Focus on Satellite Launch and In-Orbit Policies*—the Air & Space Lawyer Volume 21, No. 4, 2008; Diederiks-Verschoor I.H.P., *Financing and insurance aspects of spacecraft*, J.S.L vol. 24, Nos. 1 & 2, 1996, p. 99 et seq.

[41]Insurance coverage constitutes a third cost of the space project; Harrington A. J., *Legal and Regulatory Challenges to Leveraging Insurance for Commercial Space*—31st Space Symposium, Technical Track, Colorado Springs, Colorado, United States of America Presented on 13–14 April 2015; Sundahl, M., Financing Space Ventures in: Handbook of space law, 2015, p. 875.

[42]The first space insurance contract was concluded in 1965 for COMSAT's Early Bird satellite with coverage of pre-launch insurance and third party liability insurance, written by marine insurers; The coverage of launch and in-orbit risks began in 1968 with insuring an Intelsat fleet of satellites; See Iridium report of 2015 and Catalano Sgrosso G., *Insurance Implications About Commercial and Industrial Activities in Outer Space*—Citation: 36 Proc. On L. Outer Space 187, 1993, p. 192. Reeth van G., *Space and Insurance*, International Business Law, vol. 12, 1984, p. 127; B. Pagnanelli, *Tracking take-off of space insurance*, 2007; www.pagnanellirs.com/downloads/id281107.pdf, *accessed 27 August 2016*; Kuskuvelis I.I., *The space risk and commercial space insurance*, Space Policy, May 1993—different (stated that it covered also launch insurance).

[43]See appendix 1 to the Solvency II Directive, 2009/138/EC.

insurance suits the features of aviation third party liability insurance, due to the specific liability regime established in the LC, which makes it quite different from the liability involved in aviation transportation. Nevertheless, in each case it seems that the space insurance should be categorised as a large risk insurance,[44] being based on the definition of the large risks as included in Directive 2009/138/EC (Article 13, point 27). It refers to the types of the risks and features of the policyholder, as it is rather obvious that the latter prerequisites are met by most launch/ satellite operators,[45] but also due to the possible qualification of space risks as transport risks. Space insurance and space risks are not defined in insurance laws, and this is why space insurance contracts must rely on the space regulations and contractual practice in terms of the subject-matter of risks and its features.

When discussing the insurance of OOS, it cannot be considered in isolation from the whole risk management process, but must form an inherent part of it.[46] It should start from identifying the causes of loss, forecasting the future frequency, probability and severity of losses, creating mitigation plans, conducting cost-benefits analyses, implementing programmes for loss control. It includes mapping the risk (including the risks outlined in Sect. 2.2 of this chapter) followed by choosing the appropriate methods of risk reduction (technical analyses, quality control, etc.), risk avoidance (IOT schemes, risk mitigation) and risk transfer (insurance, or non-insurance transfer to third parties) and, finally, performing the systematic control of the outcome.[47] The potential of OOS should also be analysed from various risk management perspectives, including the assessment of how far the various risks can be managed using the generic capabilities or specific capabilities (including the in-built redundancy).[48] There are several issues to be considered and questions that are raised by insurers in this respect,[49] such as "whether OOS helps avoid insurance claims; who will benefit from it; who should pay for it; who has the authority to approve a service mission; and finally whether insurers underwrite differently a satellite that has a

[44]Catalano Sgrosso, G., *International Space Law*, 2011, p. 500.

[45]i.e. When the limits of at least two of the following criteria are exceeded: a balance-sheet total of EUR 6.2 million; a net turnover, within the meaning of Fourth Council Directive 78/660/EEC of 25 July 1978 based on Article 54(3)(g) of the Treaty on the annual accounts of certain types of companies, of EUR 12.8 million; an average number of 250 employees during financial year. More about the criteria of large risks, Kropka M., *Kolizyjnoprawna regulacja umowy ubezpieczenia w rozporządzeniu Rzym I*, 2010, pp. 128–139.

[46]For example, R. Parker, On-orbit satellite servicing—an insurer's perspective, Room 2/27/15.

[47]E. Baranoff, Enterprise and Individual Risk Management, Creative Commons 2012, p. 148, Frontiers of space risks. Natural Cosmic Hazards and Societal Challenges, ed. R.J. Wilman, Ch. J. Newman. Francis & Taylor Group 2018, pp. 28 and 148. D.M. Gerstein et al., Developing a Risk Assessment Methodology for the National Aeronautics and Space Administration, 2016, Library of Congress, p. 58 et subsq.

[48]Frontiers of space risks. Natural Cosmic Hazards and Societal Challenges, ed. R.J. Wilman, Ch. J. Newman. Francis & Taylor Group 2018, pp. 68 and 147. C. Preyssl, R. Atkins, T. Deak, Risk Management at ESA, ESA Bulletin no. 97, March 1999.

[49]Ibid.

reduced redundancy of components, but is cooperative with on-orbit servicing."[50] All those questions led to the spreading conviction that OOS has a potential for 'game-changing innovation', as most of the payments made by the insurers are due to component failures, deployment issues or expired resources (fuel, solar array/battery failure).[51]

Insurers have developed the top expertise on the space risk assessment factors. It is, however, quite probable that they will be impacted by OOS and need revision. Such predictions concern both technological factors and the legal context, as considered in Sect. 2.2 of this chapter, adding to that as well the insurance market conditions, which suffer from not improving volatility in spite of the technological progress (causing some insurers to abandon space insurance underwriting).[52] In the first stage, it will probably even increase the need for an individual approach to space insurance underwriting, which as a rule is tailored risk coverage, though in the long term it is clear that standardisation would be more than welcome.[53] The risk assessment in space insurance works on the basis of a 'technology-based engineering analysis', rather than on the typical methods of risk measurement and statistics.[54] This situation is due to the low quantity of risks of high value, i.e. the limited number of launches and satellites, which do not allow really meaningful statistics to be developed, increased by the diverse range of launch vehicles and satellites, which further narrow the possibility to act on the rules of probability.[55] This issue is related to a broader aspect

---

[50]R. Parker, On-orbit satellite servicing—an insurer's perspective, Room 2/27/15.

[51]J. Schmidt, On-orbit Satellite Servicing—Insurance Considerations, Second International Workshop on Orbit-Servicing NASA's Goddard Space Flight Center, May 2012.

[52]Traditionally, these factors include the complexity of market conditions and purely technical risk factors and among them spacecraft configuration, performance margins, track record (as well as launch vehicle track record), the insured's history. See more in AON ISB Space Insurance Fundamentals; Part I Introduction to Space Risks Management.

[53]Ibid. Space insurers distinguish several points in the space mission that are important from the risk assessment point of view. These are the intentional ignition, lift off, ascend phase and injection, satellite separation, deployment of solar panels and antennas, satellite orbit raising, satellite in-orbit testing and finally the satellite acceptance. Though only the last ones happen in orbit, there is no doubt that, unless there is a total loss during a failed launch, it seems like OOS may become a remedy for partial failures and the majority of early stage problems, from the moment the satellite is placed in orbit, even if in the incorrect one.

[54]e.g. Blassel P., *Space projects and the coverage of associated risks*, The Geneva Papers on Risk and Insurance, Vol. 10, No. 35, 1985, p. 64 "assessment of the risk of correct satellite operation over a given period of time is a complex matter, which does not lend itself exclusively to a mathematical or statistical analysis." Meredith, P., Robinson G., *Space Law: A Case Study for the Practitioner: Implementing a Telecommunications Satellite Business Concept*, Amsterdam: Martinus Nijhoff Publishers, 1992, p. 337; the individualistic approach is adopted in the case of atypical risks, where assets are of high value, for example in vessels on the high sea, Williams C.A. et al., *Risk Management and Insurance*, 2002, p. 158; Ronka-Chmielowiec W., *Ubezpieczenia Rynek i Ryzyko*, Polskie Wydawnictwo Ekonomiczne, 2002, p. 172.

[55]Kuskuvelis I.I., *The space risk and commercial space insurance*, Space Policy, May 1993, p. 111. Another factor making the underwriting endeavour difficult is limited access to data on space

of the insurability of space risks in the context of the general criteria of insurability developed by insurance practice and the doctrine.[56]

The insurance market practice divides the space risks according to the phases of the space project and distinguishes the 'launch', 'early in-orbit' and 'in-orbit' insurance.[57] This is due to the fact that risk in each subsequent phase of the space mission is very different as regards the level of exposure (the biggest being during the launch and the first two months after the launch, statistically responsible for 10% of losses in total, and for 36% of full losses).[58] The time of coverage and the scope of coverage differ as well.[59] First stage, one year up to five years are covered under the launch insurance contract. The second and the last stage covers up to 10–15 years of in-orbit life, when the satellite's value drops substantially and represents no book value. Extending the satellite's life creates a potential for further insurance coverage, along with the operative value of the satellite despite the advanced lifetime. Specialised space insurers do offer combined coverage for typical space risks, i.e. launch, early in-orbit and in-orbit. The reason for structuring

projects that were not insured, which may not have any direct influence on rates on the space insurance market, but does further limit the database for developing meaningful statistics. Finally, national statutory impediments are imposed on the transfer of data concerning space assets, the best known of which is ITAR, binding in the US. Its provisions require insurers from outside the US to obtain a licence in order to be able to see data necessary for the risk assessment, since it is recognised as an export of technical data. See also Whearty R., *Intro to Space Insurance. First party*—Marsh Space Projects a History of Leadership and innovation August 2015; Bender R.G., *International Arbitration—Satellite Communications: Arbitrator Perspective* in: International Commercial Arbitration Practice: 21st Century Perspectives, LexisNexis, 2010.

[56]Kowalewski E., *Prawo ubezpieczeń gospodarczych*, Branta 2006, p. 41 (where it is emphasised that the insurable risk should be measured by statistical methods; risk measured only with probability methods is uninsurable and the risk measured by estimations is conditionally insurable); also Kwiecień I., *Ubezpieczenie w zarządzaniu ryzykiem działalności gospodarczej*, C.H Beck, Warszawa 2010, p. 125. Kunstadter C., *Space insurance market overview*, AIAA Workshop, 2013; Space insurance market overview, Masson-Zwaan T., *Liability & Insurance in Air & Space Law; Regulation of Suborbital Flights in Europe*, Montreal, 18 March 2015, ICAO/UNOOSA Aerospace Symposium.

[57]The pre-launch coverage, until lift off, is often provided not by the space insurers, but by the cargo, marine and other insurers. This is due to the fact that all the risks related to the ground activity, even if connected with outer space have much more in common with other insurance of ultra-hazardous activities, and as such are more similar in dealing with such risks (as in the nuclear and chemical industries). Space insurance starts with the lift off of the launch vehicle and may last for the duration of the satellite's life. Kunstadter C., *Space insurance market overview*, AIAA Workshop, 2013; D. Rora, In orbit Servicing Insurance Aspects, World Risk Forum—Dubai 2010.

[58]R. Gubby, D. Wade, D. Hoffer, Preparing for the worst: The Space Insurance Market's Realistic Scenarios, New Space 2016, Vol. 4, No. 2, D. Benoussan, TeSeR—Technology for Sel-Removal of Spacecraft (Project under Horizon 2020 under grant agreement No. 687295), p. 5.

[59]The launch phase lasts no longer than one hour (depending on the type of the launch vehicle and intended orbit), the early in-orbit phase (depending on the type of the satellite may last from several weeks up to several months in the case of all-electric satellites) and the operational stage may exceed 15 years.

combined space insurance products is the difficulty in drawing a distinction between subsequent phases of the space operation, and it can be even more difficult to discover the moment when the covered risk occurred, and the moment when the loss manifests, depending on the policy wording.[60]

Apart from the above, the classical division for property (first party risk) and third party liability is made, where the latter is usually a compulsory type of insurance under a national authorisation regime and is related with the need to shift the liability borne by the state for space activities of the national space enterprises. Second party insurance, known in transport insurance as insurance of liability (to passengers or other contracting parties) does not exist in the space insurance sector, mainly due to the common application of cross-waivers in the space industry between the participants of the space missions, as well as due to the lack of regulations on second party liability in the Liability Convention.[61] Third party space liability insurance, which is compulsory, is one of the few aspects regulated in law specifically with respect to space insurance. The obligation to insure, however, does not result from international law, where the treaties are silent about insurance and are limited only to the rules of liability imposed on the launching states. In fact, only UN resolutions include suggestions on regulating the compulsory third party liability insurance in national space laws. Thus, the obligation to insure is present in the domestic space laws enacted by spacefaring states, and one of its aims is to secure the international liability of the state for space activity conducted by national entities.[62] In a number of states, the insurance obligation has not been reflected explicitly in the law, but constitutes a condition to obtain a licence to conducting space activities. The obligation to have insurance is nowadays perceived as one of the most important issues of national space legislation and is present in all the proposals concerning the harmonisation of domestic space laws.[63] It can also be noted that all newer laws on space activities include the explicit obligation to have insurance covering the risk of damage that may be sustained by a third party.[64]

---

[60]Pre-launch period (including manufacturing, testing and transportation phase, as well as the preparatory actions at the launch site), though included commonly to space insurance in a broad meaning, in fact is insured by insurers specialising in more general types of insurance, e.g. transportation, marine, or large corporations with a substantial capacity (e.g. AXA or Munich Re) based on the rules specific to all other branches of industry, even if taking into account their ultra-hazardous nature; Schöffski O., Wegener A.G., *Risk Management and Insurance Solutions for Space and Satellites Projects*, 24 The Geneva Papers on Risk and Insurance, 1999, p. 205.

[61]Masson-Zwaan T., *Liability & Insurance in Air & Space Law; Regulation of Suborbital Flights in Europe*, Montreal, 18 March 2015, ICAO/UNOOSA Aerospace Symposium; the second parties explicitly excluded from the 'space liability regime' (i.e. astronauts, passengers, etc.).

[62]Horl K. U., *Legal Aspects of Risks Involved in Commercial Space Activities*, Montreal 2003, p. 152.

[63]See, for example, Gerhard M., Schrogl K-U, *Report of the Project 2001 Working Group on National Space Legislation*, in: Project 2001 Legal Framework for the Commercial Use of Outer Space: Karl-Heinz Böckstiegel (Ed.), p. 557; and Sophia model law.

[64]See, for example, also the Polish draft space law. One of the drafts, i.e. the Sofia model law on space insurance, also includes proposed provisions on liability and compulsory liability insurance.

We may also hope that it will have a positive impact on property damage insurance by lowering the exposure for property and revenue losses. Considering the above features of space insurance contracts, as well as the nature of the OOS activity, the interaction can be seen in several contexts. Depending on the technological and legal outcomes, a new type of risk may emerge, such as second party risks (contractual liability) and changes in the third party liability paradigm into risk-based liability. The OOS may also affect certain specific features of a space insurance contract, putting the insurer more in control of risk exposure. At a minimum, there would be the possibility of the OOS vehicle inspecting the satellite, which may have an impact on loss adjustment. It will be easier to assess whether the malfunction is permanent or can be remedied. The other insurance consequence is the possibility of introducing new criteria of type of loss assessment, which lowers the risk of a total loss or TCL in terms of PD as well as in terms of BI—in view of the emerging possibility of restoring the service of the satellite by OOS spacecraft. No doubt, however, new risk will emerge, this being related to the possibility of the servicing spacecraft damaging the satellite. This will implicate the need to assess the (underwriting) exposure of PD, as well as liability (second and third party risk). In the latter case, this also needs to be addressed in OOS contracts or other arrangements (depending on which actors would be involved, private or public).

From analysing the most specific terms of current space insurance policies, it seems that the coverage of OOS could have an impact on the following: (1) damage mitigation, (2) risk assessment for the insurer at the in-orbit stage, (3) cause of damage detection in terms of recourse actions being more possible (where the liability of other operator—the owner of the damaging debris—would be involved), (4) changing the insurance paradigm from all-risk insurance to named perils insurance as a result of the above, (5) reducing the number of catastrophic losses or TCL, due both to better loss detection as well as possibility of remedying the damage. We must understand that no single solution of the above fits all, especially where different technical methods are emerging (with docking or without docking to the satellite).

As mentioned, space insurance is based on the concept of all risk, which means that all the failures and losses described and defined in the insuring contract are covered, without any reference made to the cause of such loss.[65] OOS may potentially lead to this rule changing and coverage of named perils might be proposed instead—even though all-risk insurance provides for certain exclusions, in practice space insurance can only set exclusions that can be tracked from Earth. As a result of an inspection being possible on the cause of loss, this situation may substantially change. For the same reason, the exposure assessment would be much easier, leading potentially to the change of 'agreed value' space insurance policies[66] into more conventional ones, where the calculation of the loss is based on actual circumstances, as inspected with the help of OOS spacecraft. The same is true for the precautionary measures, which

---

[65] K. Malinowska, Space Insurance, p. 287.

[66] See more, K. Malinowska, Space insurance, p. 311.

will be much easier to undertake before the loss and the mitigation measures after the insured event. Thus, with OOS implemented, loss may be avoided (for example, by the satellite achieving the proper orbit), or mitigated in such a way that partial and total constructive loss can be avoided. Finally, even only with inspection on site, the proof of loss (or to the contrary—proof of the loss due to a risk not covered). As OOS can contribute to prolonging the satellite's lifetime, it may also result in the prolongation of the insurance period of cover. Nowadays, the market practice is to limit the offered period of cover to the first years in orbit. Together with the substance of the satellite, insurance against the loss of revenues generated by satellites would make sense for a longer period. All these contribute to the stabilization of the volatile space insurance market (with the growing number of risks covered, the law of large numbers would be easier to apply, which would benefit the whole market).

Another issue that has been applied already successfully in practice and is increasingly worth considering, is for insurers to salvage the satellites. Though the salvage is problematic under space law (both international and national), under space insurance it functions upon the explicit terms of the policy.[67] It may affect a particular insurance contract in terms of the insurer's exposure, though it may also be analysed in a broader context, concerning active debris removal. Currently, there is no hard law, and not even soft law guidelines, that provide for obligation to salvage a satellite. Technical possibilities in that respect could go hand in hand with the legal obligations imposed on the operators to remove the wreck from orbit. This, in turn, may have an implication for liability and its insurance.[68] Because, as per the Outer Space Treaty, nations retain ownership of everything they put in space, current space law does not authorise the salvage or recycling of materials. This is a potential area of change in space law, though it is not expected that nations would be willing to give up ownership of their own debris in orbit. A discussion regarding salvage would be useful from both regulatory and industry perspectives.[69]

Space TPL insurance, as with any other liability insurance, is strictly correlated with the legal liability regime applicable to a given venture. This is true for the space industry as any other, and will also be true for TPL insurance related to OOS. That is why the analysis of liability insurance always starts from presenting the liability regime. Having said that, the present status quo in this respect provides for the basic regime regulated in the Outer Space Treaty and the Liability Convention. In addition, there are some national laws that aim to shift the burden of TPL to the

[67]That was already applied when the first OOS mission in the 1980s took place, allowing insurers to generate dome profits after restoring the service of the satellite Orion 3, placed primarily in an incorrect orbit. See: R. Parker, On-orbit satellite servicing—an insurer's perspective, Room 2/27/15. See more K. Malinowska, Space insurance, pp. 377–380; On-Orbit Satellite Servicing Study. Project Report, NASA Goddard Space Flight Center, October 2010, p. 80.

[68]D. Rora, In orbit Servicing Insurance Aspects, World Risk Forum—Dubai 2010.

[69]S. A. Carioscia, B. A. Corbin, B. Lal, Roundtable Proceedings: Ways Forward for On-Orbit Servicing, Assembly, and Manufacturing (OSAM) of Spacecraft, 2018 Institute for Defense Analyses.

satellite operators and to provide for a risk-sharing regime between the state and the private operators in such a way that it also usually limits the exposure of the private operators, though connected with compulsory TPL insurance or other comparable guarantee.[70]

The concept of TPL insurance is defined as the "indemnification of all sums that the insured becomes legally obliged to pay for bodily injury and/or property damage to third parties arising out of the pre-launch, launch and in-orbit operations of spacecraft." The compulsory nature of this insurance strictly depends on the shape of the domestic legal regulations (it may be partially compulsory, and in other parts not be included in the compulsory regime, depending on the phase of the mission). The highest sum insured, required by the regulators, is USD 500 million. TPL insurance may respond in the case of debris impact (assuming the owner of the debris can be identified) as well as in case of collision with an active satellite (assuming that fault is acknowledged or proven).

2.4 Conclusions

Though this Chapter tackles only general aspects of the risks and their management, the several conclusions may be drawn. Though on orbit servicing is a concept known from many years, the technological changes which have taken place during that time have an impact on its risk approach and management in the context of the business modelling and regulatory aspects. Among many, we may mention the elimination (or at least limitation) of the human factor and basing it on robotics. Also for these reasons, this sector will be more open for NewSpace ventures. This in turn implicates changing approach to the risk management processes, addressing the management differences in big, established corporations and start-ups. There are also however issues that must be consistent on global despite the business models. This concerns an integrated risk management approach including all the phases of the projects and all the stakeholders. Quite naturally there is also a place for insurers, which can provide useful tools for risk mapping, not repeating even their value in securing the whole project by insurance coverage. On orbit servicing is also a challenge for the policymakers and regulators, to apply the relevant risk management logic in the licensing requirements and liability regime (at least).

On orbit servicing poses both challenges and advantages for insurers. No doubt there will be new types of hazards and risks, the technology being unproven as regards its robustness and reliability. That can implicate the necessity of being involved in the design phase of the projects with no place for standardized approach. On the other side, the insurance industry may profit for the regulations imposing for example obligation to insure the on-orbit stage and on-orbit servicing missions as such and

[70]D. Rora, In Orbit Servicing Insurance Aspects, World Risk Forum—Dubai 2010.

possibility of involving NewSpace business in on-orbit projects may implicate more demand for insurance coverage. All this can contribute to less volatility on the space insurance market. Finally technological benefits of on-orbit servicing may enable the change of some of the paradigms of space insurance policy terms, limiting the exposure of the insurers. All of these aspects would lead to the space insurance market becoming less volatile and more accessible for more insurers and more operators.

Katarzyna Malinowska, Ph.D., Dr. hab., is an expert in insurance and space law, Professor at Kozminski University in Warsaw and the head of the Centre of Space Studies at Kozminski University (Poland). Moreover, she is an author and co-author of several books and numerous articles on insurance and space law, regular speaker at Polish and international conferences and lecturer of space, insurance and financial law. Katarzyna is an attorney-at-law, as well as the Chairperson of the Polish section of AIDA (Association Internationale de Droit des Assurances).

Chapter 3
Legal Aspects of Space Recycling

Zhuang Tian and Yangyang Cui

Abstract Space actors and practitioners have proposed various projects on space recycling, some of which have already been initiated. Based on a brief introduction of these projects, this article highlights some legal issues space scavengers may come across. The discussion is divided into four parts. Part I explores the source of law applicable to space recycling, with a distinction made between governmental and non-governmental operators. Part II analyzes the legal issues prior to mission operations, in particular the legal status of recycling targets and the transfer of jurisdiction thereover and ownership thereof. Part III examines the liability risks during mission operations, with emphasis put on the determination of and defense against fault. Part IV discusses the legal status of objects manufactured through 3D printing as it is one prospective utilization of recycled materials. Finally, some conclusions are drawn with a potential path forward sketched.

3.1 Introduction

Since the launch of Sputnik I by the Soviet Union in October 1957, about 8950 satellites have been launched into outer space, with around 5000 still in Earth orbit.[1] Among these orbiting satellites, more than half are non-functional.[2] Other types of defunct man-made objects orbiting the Earth include rocket bodies, mission-related

[1]ESA, "Space debris by the numbers" (February 2020), https://www.esa.int/Safety_Security/Space_Debris/Space_debris_by_the_numbers. (All websites cited in this article were last accessed and verified on February 27, 2020).

[2]Ibid.

Z. Tian (✉)
International Institute of Air and Space Law, Leiden University, Leiden, The Netherlands
e-mail: z.tian@law.leidenuniv.nl

Y. Cui
KoGuan Law School, Shanghai Jiao Tong University, Shanghai, China

© The Editor(s) (if applicable) and The Author(s), under exclusive license to Springer Nature Switzerland AG 2020

A. Froehlich (ed.), *On-Orbit Servicing: Next Generation of Space Activities*, Studies in Space Policy 26, https://doi.org/10.1007/978-3-030-51559-1_3

debris, and fragmentation debris.[3] All these non-functional objects are categorized as space debris, which, due to their hypervelocity, impose risks on operational satellites and human space flight. The Kessler Syndrome, a phenomenon identified by the NASA scientist Donald J. Kessler in 1978 regarding the self-sustaining cascading effect of space debris in low-Earth orbit (LEO), may deteriorate the *status quo*.[4] In this regard, a simulation modelling space debris proliferation conducted by Liou and Johnson indicates that fragments generated from mutual collisions among existing objects may force the debris population to increase over time even without any future launches.[5]

There are two traditional ways to deal with the space debris problem, namely space debris mitigation and remediation. While the former emphasizes the reduction of the increase rate of *new* pieces of space debris, the latter refers to the active removal of *existing* debris from space.[6] According to Liou's modelling results, to stabilize the space debris population, the removal of five large objects per year and a 90% success rate of post-mission disposal for all launched objects are needed.[7] This demonstrates the demand for synergy of debris mitigation and remediation, as neither of them can solve the space debris problem alone.

A third option to reduce space debris is on-orbit servicing, which aims at the extension of operational lifetime of satellites and the refunctionalization of defunct objects in outer space. As indicated by its name, on-orbit servicing is the technology to service objects which have already been launched in space.[8] In a project conducted by Team DOCTOR of the International Space University, it is defined as follows:

> On-Orbit Servicing is a service offered for scientific, security, or commercial reasons that entail an in-space operation on a selected client spacecraft to fulfill one (or more) of the following goals: inspect, move, refuel, repair, recover from launch failure, or add more capability to the system.[9]

---

[3]NASA, "Orbital Debris Quarterly News", Volume 24 - Issue 1 (January 2020), 4.

[4]Donald J. Kessler and Burton G. Cour-Palais, "Collision frequency of artificial satellites: The creation of a debris belt", *Journal of Geophysical Research: Space Physics* 83, no. A6 (1978), 2637–2646.

[5]J. C. Liou and Nicholas L. Johnson, "Risks in Space from Orbiting Debris", *Science* (Washington) 311, no. 5759 (2006), 340.

[6]UN Doc. A/AC.105/C.1/2012/CRP.16, "Active Debris Removal—An Essential Mechanism for Ensuring the Safety and Sustainability of Outer Space" (January 27, 2012), 7.

[7]J. C. Liou, "Engineering and Technology Challenges for Active Debris Removal", 4th European Conference for Aerospace Science (July 4–8, 2011), 5–6.

[8]Mahashreveta Choudhary, "What is on-orbit satellite servicing?" (October 15, 2019), https://www.geospatialworld.net/blogs/on-orbit-satellite-servicing-process-benefits-and-challenges/. Alexander Soucek, "On-Orbit Servicing: Legal Perspective", European Space Policy Institute (April 18, 2018).

[9]Team DOCTOR, "Developing On-Orbit Servicing Concepts, Technology Options, and Roadmap", Final Report of the International Space University Summer Session Program 2007, 1. https://isulibrary.isunet.edu/doc_num.php?explnum_id=102.

On-orbit servicing entails various mission types, including repairing, refueling, upgrading, re-orbiting, and recycling.[10] The legal analysis of the last type, namely the recycling of objects in outer space, is the theme of this article. As there is no legally binding definition of "space recycling", its meaning should be explored elsewhere. In its ordinary meaning, recycling refers to "the action or process of converting waste into reusable material".[11] To put it in the space context, "space recycling" generally means the scavenging of space debris and the subsequent reutilization thereof. The success of 3D printing in the International Space Station (ISS) demonstrates the prospect of space recycling. In November 2014, the ISS's 3D printer produced the first 3D printed object in space, starting a new chapter of off-Earth manufacturing.[12] In November 2018, the first integrated 3D printer and recycler known as "Refabricator" was launched to the ISS.[13] Compared to its predecessor, the Refabricator moves one step forward because it entails the capability to reprint multiple times.[14]

Space actors and practitioners have put forward several proposals on space recycling. Gateway Earth Development Group, a collection of academics from universities across the world, proposes to turn space debris into useful resources.[15] Their plan is to place Gateway Earth, a fully operational space station, in geostationary orbit (GEO) to recycle defunct satellites and other categories of space debris.[16] Tom Markusic, CEO of Firefly Aerospace, also proposes to recycle and reuse "dead satellites" in GEO.[17] His idea is to tow these objects beyond Earth orbit and reutilize them as supporting equipment for missions to Mars.[18]

The recycled materials may be used to repair broken satellites and to build new ones. U.S. Defense Advanced Research Projects Agency (DARPA) started a project called Phoenix in 2011 which planned to launch a "spacecraft that could remove a solar array, antenna, or other component from a defunct satellite and transport it to another satellite, either a newly-constructed spacecraft or one in need of repairs".[19]

---

[10] Thea Flem Dethlefsen, "On-Orbit Servicing: Repairing, Refuelling and Recycling The Legal Framework", 70th International Astronautical Congress (IAC), Washington D.C., United States, 21–25 October 2019, 1–2.

[11] Oxford University Press, https://www.lexico.com/.

[12] NASA, "International Space Station's 3-D Printer" (November 26, 2014), https://www.nasa.gov/content/international-space-station-s-3-d-printer.

[13] Jennifer Harbaugh, "Combination 3D Printer will Recycle Plastic in Space" (November 19, 2018), https://www.nasa.gov/mission_pages/centers/marshall/combination-3d-printer-will-recycle-plastic-in-space.html.

[14] Korey Haynes, "The Space Station's New 3-D Printer Recycles Old Plastic Into Custom Tools" (February 11, 2019), https://www.discovermagazine.com/the-sciences/the-space-stations-new-3-d-printer-recycles-old-plastic-into-custom-tools.

[15] Jez Turner, "What you gon' do with all that junk? Why space needs a recycling station" (August 7, 2019), https://www.smartcompany.com.au/startupsmart/analysis/space-junk-recycling-station/.

[16] Ibid.

[17] Eric Mack, "The Company Betting That Space Junk Could Be a Big Business" (February 25, 2016), https://www.inc.com/eric-mack/theres-a-plan-to-recycle-space-junk-at-mars.html.

[18] Ibid.

[19] Jeff Foust, "The space industry grapples with satellite servicing" (June 25, 2012), https://www.thespacereview.com/article/2108/1.

DARPA revamped the Phoenix Project in 2015 to shift its focus on the servicing of geostationary satellites.[20] Based on the Phoenix Project, in 2016 DARPA launched a new public-private partnership program called Robotic Servicing of Geosynchronous Satellites (RSGS). The RSGS program seeks to develop technologies that would enable the inspection and servicing of geosynchronous satellites.[21]

In addition to the recycling of satellites, ideas have also been conceived on the repurposing of rocket bodies. The U.S. company Nanoracks is partnering with the Canadian Company Maritime Launch Service (MLS) to recycle the latter's rocket stages for the creation of "outposts" in outer space.[22] These rockets are planned to be launched in the fall of 2021 and would be subsequently transformed into storage platforms of satellites or fuel depots.[23] According to Jeffrey Manber, CEO of Nanoracks, the first phase of the project is to repurpose rocket stages, and the second phase to examine the potential to reutilize those pieces of debris already in space that are larger than 10 centimeters.[24]

As shown in the cases above, space recycling is evolving from a concept to reality. There are two major benefits of space recycling. First, harvesting materials from defunct satellites and rocket bodies is potentially much less expensive than transporting materials from Earth.[25] Its potential to cut costs would constitute an economic incentive for operators to invest in space recycling technologies. Moreover, space recycling helps to alleviate the crowdedness in outer space as the repurposing and reutilization of defunct objects in orbit will reduce the need to launch new objects from Earth.

This article aims to highlight some key legal issues the operators of space recycling ("space scavengers") should take into consideration before entering this promising and dynamic domain, which include the identification of applicable law, the legal status of the recycling targets, the liability issues, and the legal status of objects manufactured from recycled materials.

---

[20] Caleb Henry, "DARPA Revamps Phoenix In-Orbit Servicing Program" (June 2, 2015), https://www.satellitetoday.com/government-military/2015/06/02/darpa-revamps-phoenix-in-orbit-servicing-program/.

[21] DARPA, "Program Aims to Facilitate Robotic Servicing of Geosynchronous Satellites" (March 25, 2016), https://www.darpa.mil/news-events/2016-03-25.

[22] Frances Willick, "Canso spaceport partners with U.S. company to recycle rockets in space" (November 3, 2019), https://www.cbc.ca/news/canada/nova-scotia/mla-nanoracks-repurpose-cyclone-4m-rockets-1.5344086.

[23] Keith Doucette, "Nova Scotia spaceport teams with U.S. firm to examine reuse of rockets in space" (November 4, 2019), https://atlantic.ctvnews.ca/nova-scotia-spaceport-teams-with-u-s-firm-to-examine-reuse-of-rockets-in-space-1.4669438.

[24] Ibid.

[25] Eric Mack, "The Company Betting That Space Junk Could Be a Big Business" (February 25, 2016), https://www.inc.com/eric-mack/theres-a-plan-to-recycle-space-junk-at-mars.html.

3.2 The Applicable Law

The legal framework of space activities consists mainly of treaties, non-legally binding instruments released by the United Nations (UN) and other international organizations, and the national legislation of more than 20 countries.[26] The core of this framework are the five international treaties developed under the auspices of the UN between 1967 and 1979, namely the Outer Space Treaty,[27] the Rescue Agreement,[28] the Liability Convention,[29] the Registration Convention,[30] and the Moon Agreement[31] (collectively as the "UN space treaties"). The Outer Space Treaty, often termed as the Magna Carta of international space law,[32] provides basic principles for the regulation of space activities, which have been reiterated, elaborated, and developed in the subsequent four treaties.

Under general international law, the conduct of any State organ shall be considered an act of its national State.[33] Therefore, governmental operators shall assure that their activities are carried out in conformity with international law, the breach of which will give rise to an international responsibility of its national State.[34]

For non-governmental operators, their concern lies primarily in national space legislation which directly regulates their rights and obligations in the operation of space activities. Most commonly, States tend to extend the scope of their national

---

[26]Rada Popova and Volker Schaus, "The Legal Framework for Space Debris Remediation as a Tool for Sustainability in Outer Space", *Aerospace* 5, no. 2 (2018), 55, 4. The United Nations Office for Outer Space Affairs (UNOOSA), "Space Law", https://www.unoosa.org/oosa/en/ourwork/spacelaw/index.html.

[27]Treaty on Principles Governing the Activities of States in the Exploration and Use of Outer Space, Including the Moon and Other Celestial Bodies; 610 UNTS 205; entered into force on October 10, 1967.

[28]Agreement on the Rescue of Astronauts, the Return of Astronauts and the Return of Objects Launched into Outer Space; 672 UNTS 119; entered into force on December 3, 1968.

[29]Convention on International Liability for Damage Caused by Space Objects; 961 UNTS 187; entered into force on September 1, 1972.

[30]Convention on Registration of Objects Launched into Outer Space; 1023 UNTS 15; entered into force on September 15, 1976.

[31]Agreement Governing the Activities of States on the Moon and Other Celestial Bodies; 1363 UNTS 13; entered into force on July 11, 1984.

[32]Qizhi He, "The outer space treaty in perspective", *J. Space L.* 25 (1997), 93.

[33]International Law Commission (ILC), Responsibility of States for Internationally Wrongful Acts (2001), Article 4(1). The customary character of this rule has been confirmed by the International Court of Justice (ICJ). See *Difference Relating to Immunity from Legal Process of a Special Rapporteur of the Commission on Human Rights, Advisory Opinion, I.C.J. Reports 1999*, p. 62, at p. 87, para. 62.

[34]Ibid., Article 1 and 2.

space legislation to activities carried out from their (quasi-)territory and to activities conducted by their nationals regardless of the venue. Therefore, space scavengers shall look into both *lex patriae* and *lex loci actus*, namely the law of nationality and the law of place of performance.[35] Without a universal consensus on the meaning of "national activities" and "appropriate State" under Article VI of the Outer Space Treaty, States have the discretion to define the scope of application of their national space legislation. Therefore, the applicable national law for non-governmental operators should be determined on a case-by-case basis.

However, it would be advisable for non-governmental operators to look beyond national law. A special rule of attribution is established in the Outer Space Treaty, which prescribes that States shall bear international responsibility for national activities in outer space, whether such activities are carried on by governmental agencies or by non-governmental entities, and for assuring that national activities are carried out in conformity with the Outer Space Treaty.[36] The Outer Space Treaty further imposes an obligation on States to authorize and continually supervise activities of non-governmental entities in outer space.[37] In most cases, States discharge these obligations by enacting national legislation. In its Resolution 59/115, the UN General Assembly also recommends States enact and implement national laws to authorize and continuously supervise the activities in outer space of non-governmental entities under their jurisdictions.[38] The coherence between international space law and national space legislation can be therefore expected. As a result, the international legal regime is of relevance to both governmental and non-governmental operators.

3.3 Legal Status of the Recycling Targets

Due to the strategic importance, technical sensitivity, and underlying security concerns of many space objects, the first question for space scavengers is which State exercises jurisdiction over their recycling targets. The key lies in the first sentence of Article VIII of the Outer Space Treaty, which provides that the State on whose registry an object launched into outer space is carried shall retain jurisdiction and control over such object. Three related questions then arise: (a) What is "object launched into outer space"; (b) Which State is entitled to register; (c) How to understand "jurisdiction and control".

As illustrated in the Introduction, recycling targets are defunct satellites, launch vehicles, and potentially debris larger than 10 cm. These objects fall within

[35] Annette Froehlich and Vincent Seffinga, *National Space Legislation: A Comparative and Evaluative Analysis*, Studies in Space Policy, v. 15 (2018), 137–142.

[36] Outer Space Treaty, Article VI.

[37] Ibid.

[38] UN Doc. A/RES/59/115, Application of the concept of the 'launching State' (January 25, 2005).

the tracking capability of the U.S. Space Surveillance Network, one of the most advanced space surveillance systems in the world, which tracks and maintains in catalog objects down to about 10 cm in LEO and about 1 m in GEO.[39] For the cataloged objects, their launch date, launch site, and State of origin are generally recorded.[40] In other words, the launching State and the owner of a certain recycling target can be normally identified.

3.3.1 Object Launched into Outer Space

The terminology used in the UN space treaties is not consistent, which includes "objects", "objects in outer space", and "space object".[41] As these terms are frequently used interchangeably in these treaties, von der Dunk concludes that all these terms point to the same concept.[42] For the purpose of consistency, this article uses the term "space object" to analyze the legal status of recycling targets.

The Liability Convention and the Registration Convention provide that "space object" includes component parts of a space object as well as its launch vehicle and parts thereof".[43] This definition indicates what "space object" includes, but its circular wording fails to specifically define the nature, scope, and meaning of "space object". Nonetheless, a fairly precise idea on the attributes of "space object" can be inferred following the approach of treaty interpretation set by the *Vienna Convention on the Law of Treaties*[44] (the "Vienna Convention").[45]

According to the Vienna Convention, a treaty shall be interpreted in good faith in accordance with the ordinary meaning to be given to the terms of the treaty in their context and in the light of its object and purpose.[46] The ordinary meaning of "object" is a "material thing", and any "tangible thing, visible, or capable of discernment by the senses".[47] In the same vein, Mr. Kempf defines an "object" as "any tangible,

---

[39] Mark Matney, "Measuring Small Debris-What You Can't See Can Hurt You", https://ntrs.nasa.gov/archive/nasa/casi.ntrs.nasa.gov/20160011226.pdf.

[40] Loretta Hall, "The History of Space Debris" (2014), Space Traffic Management Conference, 19, https://commons.erau.edu/stm/2014/thursday/19/.

[41] Frans G. von der Dunk, "Defining Subject Matter under Space Law: Near Earth Objects versus Space Objects" (2008), *Space, Cyber, and Telecommunications Law Program Faculty Publications*, 25, 294.

[42] Ibid., 299–300.

[43] Article I(d), Liability Convention. Article I(b), Registration Convention.

[44] United Nations, Vienna Convention on the Law of Treaties, 23 May 1969, United Nations, Treaty Series, vol. 1155, p. 331, entered into force on January 27, 1980.

[45] von der Dunk, *supra* note 41, 294.

[46] Vienna Convention, Article 31(1).

[47] Ballentine's Law Dictionary, https://www.citizenlaw.com/pdf/o.pdf.

physical thing".[48] Therefore, intangible substances such as signals and emissions are excluded from "space object".

From a contextual perspective, the term "space object" equates with the term "object launched into outer space". The wording "objects launched into outer space" as used in the title and the preamble of the Registration Convention echoes the term contained in Article VIII of the Outer Space Treaty. Meanwhile, the Registration Convention uses the term "space object" throughout its provisions without any reference to "objects launched into outer space". It can thus be inferred that these two terms are interchangeable. The qualifying phrase "launched into outer space" is not to classify space objects (into launched and non-launched categories) but to describe its attributes.

With regard to the purpose, the whole teleological context of the Outer Space Treaty is to deal with activities of mankind and "man's entry into outer space".[49] Space objects' attribute of being "launched into outer space", together with the reference to "component parts" thereof, "more or less suggesting a process of 'composition' by human hands".[50] Therefore, a borderline should be drawn between man-made and natural objects.[51] The latter objects can be arguably categorized as "celestial bodies" under the UN space treaties.[52]

As each functional object launched into outer space will become non-functional over a certain period of time,[53] a related question is whether functionality is of relevance to be qualified as a space object.[54] The answer is arguably negative, for the UN space treaties do not touch upon the functionality of space objects. Moreover, excluding non-functional objects from the scope of "space object" will lead to various loopholes in the current legal framework. For instance, the State of registry would no longer retain jurisdiction and control over their registered objects after mission completion. Moreover, this exclusive approach contravenes the victim-oriented philosophy of the Liability Convention. Therefore, as commented by Masson-Zwaan, "an inactive satellite or even a lost screwdriver should still be regarded as (a component part of) a space object".[55]

---

[48] United States. Congress. House. Committee on the Judiciary. Subcommittee on Courts, I. Property. (1990). *Patents in space: hearing before the Subcommittee on Courts, Intellectrual Property, and the Administration of Justice of the Committee on the Judiciary, House of Representatives, One Hundred Frist Congress, first session, on H.R. 2946, Patents in Space Act, October 4, 1989.* Washington: U.S. G.P.O., 25.

[49] von der Dunk, *supra* note 41, 294–295.

[50] Ibid., 296.

[51] Lesley-Jane Smith and Armel Kerrest, "Article I (Definitions) LIAB", in Hobe, Schmidt-Tedd, and Schrogl (eds), *Cologne Commentary on Space Law. Vol. 2: Rescue Agreement, Liability Convention, Registration Convention, Moon Agreement*, Köln: Heymann, 2013, 115.

[52] von der Dunk, *supra* note 41, 295–296.

[53] Smith and Kerrest, "Article I (Definitions) LIAB", *supra* note 51, 115.

[54] Ibid.

[55] Tanja Masson-Zwaan, "Legal Aspects of Space Debris. International Academy of Astronautics", *IAA Space Debris Situation Report 2016*, 142.

In short, space objects are all man-made objects launched into outer space, including its component parts as well as its launch vehicle and parts thereof, which are physical and tangible. Functionality is not a criterion for its determination. Therefore, recycling targets, whether defunct satellites and rockets or smaller items and components, should all be regarded as "space object" over which their State of registry retains jurisdiction and control.

Meanwhile, these defunct objects in outer space can also be defined as "space debris" under the *Space Debris Mitigation Guidelines of the Committee on the Peaceful Uses of Outer Space* (UNCOPUOS SDM Guidelines),[56] which was endorsed by the UN General Assembly in its resolution 62/217.[57] Space debris is defined in the UNCOPUOS SDM Guidelines as "all man-made objects, including fragments and elements thereof, in Earth orbit or re-entering the atmosphere, that are non-functional".[58] However, to be categorized as space debris does not necessarily change the legal status of space objects as accorded under the UN space treaties. The UNCOPUOS SDM Guidelines is not legally binding under international law and is of voluntary nature.[59] Moreover, the Guidelines states that "exceptions to the implementation of individual guidelines or elements thereof may be justified, for example, by the provisions of the United Nations treaties and principles on outer space".[60] Therefore, in case of any conflicts between the provisions of these instruments, the UN space treaties shall prevail.

3.3.2 State of Registry

The registration obligation under the Outer Space Treaty is elaborated in the Registration Convention, which provides that when a space object is launched into outer space, the launching State shall register such object by means of an entry in an appropriate registry which it shall maintain.[61] Such launching State is termed as the "State of registry".[62] The Registration Convention spells out four categories of launching States, namely the State which launches or procures the launching of a space object,

---

[56]See https://www.unoosa.org/pdf/publications/st_space_49E.pdf. The UN Space Debris Mitigation Guidelines are based on the technical content and the basic definitions of the Inter-Agency Space Debris Coordination Committee (IADC) Space Debris Mitigation Guidelines. Other organizations, such as the International Organization for Standardization (ISO) and the International Telecommunications Union (ITU) also establish standards and recommendations with regard to space debris mitigation.

[57]UNCOPUOS SDM Guidelines, iv.

[58]Ibid., 1.

[59]Ibid., iv and 2.

[60]Ibid., 2.

[61]Registration Convention, Article II(1).

[62]Ibid., Article I(c).

and the State from whose territory or facility a space object is launched.[63] The contents of each registry and the conditions under which it is to be maintained are left to the discretion of the State of registry.[64] Meanwhile, although the Registration Convention also requires the UN General Assembly to maintain a Register to record the information of space objects as provided by the State of registry,[65] this international Register does not play any role in jurisdiction and control which is determined solely on the basis of the national registry.[66]

The obligation for the launching State to register space objects also raises a question on the legal status of unregistered space objects. Schmidt-Tedd and Mick are of the view that "without the first step of national registration, no jurisdiction and control over the space object in question is feasible".[67] In contrast, Palkovitz argues that apart from a mere non-compliance of treaty obligations for State Parties to the Registration Convention, there is no difference between a registered satellite and an unregistered one.[68] The terminology "retain" which literally means "continue to have" and "keep possession of" appears to incline to the latter interpretation.[69] It indicates that the act of registration does not in itself create but serves as recognition of pre-existing jurisdiction and control. Moreover, the responsibility to authorize and supervise national space activities under Article VI of the Outer Space Treaty entails a de facto exercise of jurisdiction over space objects. Therefore, unregistered objects are not void of jurisdiction and control.

3.3.3 Jurisdiction and Control

Under general international law, jurisdiction is an aspect of sovereignty, which "refers to a State's competence under international law to regulate the conduct of natural and juridical persons".[70] Interpretation of the term "control" is somehow divergent. While some commentators regard it as the "factual element which ensures that the

[63] Ibid., Article I(a). The same definition can be found in Article I(c) of the Liability Convention. The latter Convention further clarifies in Article I(b) that the term "launching" includes attempted launching.

[64] Registration Convention, Article II(3).

[65] Ibid., Article III(1).

[66] Belgium, Response to the Questionnaire, UN Doc. A/AC.105/C.2/2012/CRP.11 (March 22, 2012), 4.

[67] Bernhard Schmidt-Tedd and Stephan Mick, "Article VIII", in Hobe, Schmidt-Tedd, and Schrogl (eds), *Cologne Commentary on Space Law. Vol. 1: Outer Space Treaty.* Köln: Heymann, 2009, 152.

[68] Neta Palkovitz, "Regulating a Revolution: Small Satellites and the Law of Outer Space", Ph.D. Dissertation, Leiden University, 2019, 176–177.

[69] Oxford Dictionary Online, https://www.lexico.com/en/definition/retain.

[70] Ian Brownlie and James R Crawford, *Brownlie's Principles of Public International Law*, 8th ed, Oxford University Press (2012), 456.

possibility to technically control the satellite lies within the State registry",[71] others define it as the "exclusive right and the actual possibility to supervise the activities of a space object" which "must be based on legitimate jurisdiction and not on factual control capabilities".[72] Since the State of registry withholds jurisdiction and control over a space object even after its end-of-mission, "the 'control' competence is more than a technical capability".[73]

A related question is whether the term "jurisdiction and control" entails rights or obligations of the State of registry. Arguably, it refers to both. A right can be inferred as the term "jurisdiction" itself is an aspect of sovereignty. Moreover, interpreting this term as a pure obligation will lead to a manifestly absurd and unreasonable result as States would then lose their interest to register space objects. Meanwhile, if this term is interpreted in a restricted sense of entailing merely rights without corresponding obligations, the underlying purpose of Article VI of the Outer Space Treaty would be undermined.[74] In its response to the *Set of Questions provided by the Chair of the Working Group on the Status and Application of the Five United Nations Treaties on Outer Space* ("Questionnaire"), Belgium submits that "the exercise of such jurisdiction and control entails prerogatives but also obligations".[75] Therefore, if a State intends to recycle a space object, it is entitled to do so only if it has jurisdiction and control over the recycling target or the permission from the State of registry of such target is granted.[76]

As mentioned in the Introduction, U.S. company Nanoracks plans to recycle Canadian company MLS's rockets to create outposts in outer space. In this case, Nanoracks can be regarded as somehow "procuring" the launch of MLS's rocket as they have negotiated the recycling issues before launch. Therefore, one may argue that the U.S. is a launching State of MLS's rockets, and hence the registration of the rocket bodies as well as jurisdiction and control retained thereover can be transferred from Canada to the U.S. Notedly, without a clear definition of "procure" given in the UN space treaties, whether the pre-launch negotiation can be considered as "procure" remains questionable. Moreover, whether a State can be categorized as a launching State when its nationals procure the launching of a space object is also debatable.

If the State which bears international responsibility for a space scavenger's space activities under Article VI of the Outer Space Treaty ("responsible State") is not the launching State of a recycled object, the question regarding the exercise of jurisdiction arises. As previously mentioned, under the Registration Convention only launching States are entitled to register space objects. This means that the transfer of registration

---

[71] Popova and Schaus, *supra* note 26, 9.

[72] Schmidt-Tedd and Mick, "Article VIII", *supra* note 67, 157.

[73] Ibid.

[74] H Bittlinger, Hoheitsgewalt und Kontrolle im Weltraum (Carl Heymanns, Cologne 1988) 33; G Gal, Space Law (Sijthoff, Leiden 1969) 215. Citing from Schmidt-Tedd and Mick, "Article VIII", *supra* note 67, 158.

[75] Belgium, Response to the Questionnaire, UN Doc. A/AC.105/C.2/2012/CRP.11 (March 22, 2012), 4.

[76] UN Doc. A/AC.105/C.1/2012/CRP.16, *supra* note 6, 32.

of space objects can only be made amongst the launching States, while the transfer from the State of registry to a non-launching State seems impossible. As jurisdiction and control is determined on the basis of registration, this places a legal constrain on space recycling activities. As in the Nanoracks case, if the U.S. could not be categorized as a launching State, it would not be eligible to register the rockets even after the transfer of ownership between MLS and Nanoracks.

To solve this problem, the responsible State may enter into an agreement with the State of registry to deal with jurisdictional issues. Reference can be made to the Article II paragraph 2 of the Registration Convention which provides that

> Where there are two or more launching States in respect of any such space object, they shall jointly determine which one of them shall register the object [...], bearing in mind the provisions of article VIII of the *Outer Space Treaty*, and without prejudice to appropriate agreements concluded or to be concluded among the launching States on jurisdiction and control over the space object and over any personnel thereof.

The last part of this provision implies that the exercise of jurisdiction and control does not rest upon the State of registry in an absolute manner, and therefore, specific arrangements can be made by launching States. It does not impose any requirement nor restriction on the content of such arrangements. As a result, launching States have the freedom in this regard including the assignment of jurisdiction and control over space objects to a third State. Meanwhile, according to the Vienna Convention, a treaty can create both rights and obligations to a third State with its consent.[77] Bearing in mind that jurisdiction and control entails both rights and obligations, an arrangement to transfer all these to a third State is possible if it assents thereto.

It should be also noted that the responsible State, even without being the State of registry, bears international responsibility to authorize and supervise the operation of space objects under Article VI of the Outer Space Treaty. Therefore, in the event of cross-border transfer of ownership of a space object, the responsible State and the State of registry have concurrent jurisdiction over such object.[78] These two States may enter into an ad hoc agreement to arrange issues regarding jurisdiction and liability.[79]

3.3.4 Transfer of Ownership

Besides, proprietary matters also need to be solved for the operation of space recycling, especially when non-governmental operators are concerned. As prescribed in the Outer Space Treaty, ownership of space objects is not affected by their location.[80]

[77]Vienna Convention, Article 35 and 36.

[78]The Netherlands, Response to the Questionnaire, UN Doc. A/AC.105/C.2/2012/CRP.11 (March 22, 2012), 8.

[79]Ibid.

[80]Outer Space Treaty, Article VIII.

Therefore, space scavengers should enter into a contractual agreement with owners of target objects for the transfer of ownership thereof.

States normally impose authorization requirements on the transfer of ownership in their national legislation.[81] A model provision is that "the transfer of a space activity and or a space object to another operator is subject to prior authorization by the competent authority".[82] This model is adopted in some national space legislation. For instance, the Danish *Outer Space Act* prescribes that the transfer of space objects or space activities to another owner or operator may only take place after prior approval from the competent authority.[83] It further provides that if the transfer is made to another owner or operator domiciled in another State, the competent authority may impose requirements for an advance agreement with that State to take over the liability to pay damages.[84] Therefore, non-governmental operators have to go through the authorization process for the transaction.

Moreover, export control regulations may also restrict the transfer of ownership to a foreign operator. One example is the Chinese *Regulations on Control of Military Products Export* and the corresponding *Military Products Export Control List* ("Military Products Export Control Regulations and List") which stipulate export control rules for rockets, missiles, military satellites and their auxiliary equipment.[85] To further strengthen export control of missiles and missile-related items and technologies, the Chinese State Council promulgated the *Regulations of the People's Republic of China on Export Control of Missiles and Missile-related Items and Technologies* and the *Missiles and Missile-related Items and Technologies Export Control List* ("Missiles Export Control Regulation and List").[86] The latter List includes items like rockets, unmanned air vehicles, missiles and missile-related items and technologies.[87] The export of civil satellites has not been specifically stipulated in law, but the State Administration for Science, Technology and Industry for National Defense takes charge of its authorization in practice.[88] Space scavengers should look into these export control regulations and lists prior to their mission operations.

---

[81] UNGA Resolution, UN Doc. A/RES/68/74, on Recommendations on national legislation relevant to the peaceful exploration and use of outer space (December 16, 2013).

[82] ILA Space Law (2012) Resolution no. 6/2012, Annex Sofia Guidelines for a Model Law on National Space Legislation, Article 9.

[83] Danish Outer Space Act, Article 15(1). Unofficial English translation see: https://ufm.dk/en/legislation/prevailing-laws-and-regulations/outer-space/outer-space-act.pdf.

[84] Ibid., Article 15(2).

[85] Yun Zhao, "Regulation of Space Activities in the People's Republic of China", in Ram S. Jakhu, *National Regulation of Space Activities*, Space Regulations Library; Vol. 5. 238232972. New York, NY [etc.]: Springer, 2010, 263–264. The Chinese Text of the Military Products Export Control Regulations and List see https://www.fmprc.gov.cn/web/ziliao_674904/tytj_674911/tyfg_674913/t10493.shtml and https://www.chinesemission-vienna.at/chn/dbtyw/fks/c3/t206373.htm, respectively.

[86] The Chinese text of the Missiles Export Control Regulation and List see https://www.gov.cn/gongbao/content/2002/content_61742.htm.

[87] Zhao, *supra* note 85, 264.

[88] Guozhu Gao, "Current Situation, Legal Issues and Relevant Space Legislation of China's International Cooperation in Space", *Journal of Beijing University of Aeronautics and Astronautics Social Sciences Edition*, 2019, 32(5): 103.

3.4 Liability

The Outer Space Treaty sets forth that the launching State of a space object is liable for damage caused to another State or its nationals by such object.[89] This rule of liability is elaborated in the Liability Convention, under which only States (as compared to non-governmental entities) may be held liable for damage caused by a space object. Nonetheless, the States may recourse against the non-governmental operators for compensation paid to third parties, if their national legislation so provides. Therefore, non-governmental operators also have the interest to assure that their activities do not trigger international liability under the Liability Convention.

The Liability Convention establishes absolute liability for damage caused by a space object on the surface of the Earth or to aircraft in flight.[90] Zafren commented that despite the terminology "absolutely" used, the Liability Convention indeed provides for strict liability since the Convention allows a launching State to be exonerated from liability in certain instances,[91] more specifically, the contributory negligence of victims.[92]

For damage caused in outer space, a fault-based liability regime applies that the launching State shall be liable only if the damage is due to its fault or the fault of persons for whom it is responsible.[93] The Liability Convention, however, does neither define the term "fault" nor establish a standard of care for operators, which raises controversies as to its interpretation.[94] Generally, fault denotes a non-compliance or breach of an obligation imposed by law.[95] In this regard, in its response to the Questionnaire, the Netherlands states that the failure to adhere to soft law instruments cannot be phrased in terms of "fault".[96] Meanwhile, Smith and Kerrest are of the opinion that notwithstanding their non-compulsory nature, the relevant guidelines and codes of conduct may be considered by judges as evidence of the correct procedure to be followed, particularly in the context of Art. IX of the Outer Space Treaty.[97] They further argue that there will be no rooms for presumptions of fault where those

[89]Outer Space Treaty, Article VII.

[90]Liability Convention, Article II.

[91]Daniel Hill Zafren, *Convention on International Liability for Damage Caused by Space Objects: Analysis and Background Data: Staff Report*, US Government Printing Office, Volume 8, 1972, 25–26.

[92]Liability Convention, Article VI.

[93]Ibid., Article III.

[94]Joel A Dennerley, "State Liability for Space Object Collisions: The Proper Interpretation of 'Fault' for the Purposes of International Space Law", *European Journal of International Law*, Volume 29, Issue 1, February 2018, 282–283.

[95]Smith and Kerrest, "Article III (Fault Liability) LIAB", *supra* note 51, 132.

[96]The Netherlands, Response to the Questionnaire, UN Doc. A/AC.105/C.2/2012/CRP.11 (March 22, 2012), 7.

[97]Smith and Kerrest, "Article III (Fault Liability) LIAB", *supra* note 51, 133. Article IX of the Outer Space Treaty imposes a due regard obligation on States for their activities in outer space.

widely accepted guidelines and standards have been complied with.[98] Considering the risks entailed in space activities and the ever-growing space debris population, the latter interpretation corresponds more closely to the "common interest of mankind". Therefore, to reduce the risk of being held at fault, it would be in the interest of space scavengers to assure that their operations adhere to international guidelines and standards, such as the UNCOPUOS SDM Guidelines.

It has observed that fault liability may disincentivize operators to carry out on-orbit servicing.[99] In this regard, a proposal has been advanced to enact a provision, perhaps in a protocol to the Liability Convention, that for those operations which facilitate the preservation of outer space environment, fault could be mitigated in some way.[100] However, due to the very strategic and utilitarian nature of space activities, the political will of States to adopt international binding law in this regard appears to be gloomy.[101] Therefore, a protocol to the Liability Convention does not seem plausible in the short term. Alternatively, this article argues that space scavengers may refer to the *Guidelines for the Long-term Sustainability of Outer Space Activities* (the "Long Term Sustainability Guidelines")[102] as a defense against fault.

The Long Term Sustainability Guidelines was adopted by the UNCOPUOS in June 2019, with the purpose to assist States and international intergovernmental organizations to "mitigate the risks associated with the conduct of outer space activities so that present benefits can be sustained and future opportunities realized".[103] To achieve this aim, one guideline is to "investigate and consider new measures to manage the space debris population in the long term".[104] These new measures could include, inter alia, "methods for the extension of operational lifetime, novel techniques to prevent collision with and among debris and objects with no means of changing their trajectory".[105] The operation of space recycling is in line with these aims and its contribution to the preservation of outer space environment for future generations should be taken into consideration by judges to determine "fault".

Notedly, the remedies provided for under the Liability Convention are not exclusive. Rather, it leaves the domestic forum open to victims,[106] for it does not prevent

---

[98] Ibid.

[99] A/AC.105/C.1/2012/CRP.16, *supra* note 6, 32.

[100] Ibid.

[101] Ram Jakhu, Steven Freeland, and Kuan-Wei Chen, "The sources of international space law: revisited", *Journal of Air and Space Law*, 67(4), 2018, 666.

[102] UN Doc. A/AC.105/C.1/L.366, Guidelines for the Long-term Sustainability of Outer Space Activities (July 17, 2018).

[103] Ibid., 2.

[104] Long Term Sustainability Guidelines, Guideline D.2.

[105] Ibid., para. 3.

[106] Smith and Kerrest, "Article III (Fault Liability) LIAB", *supra* note 51, 133.

the victim State or its nationals "from pursuing a claim in the courts or administrative tribunals or agencies of a launching State".[107] For instance, victims may claim damages against operators on the basis of ordinary tort law.[108] Therefore, space scavengers should also consider the criterion of fault under national legislation.

3.5 3D-Printing

In addition to maintenance, refurbishing, and repurposing, recycled objects may also be used as 3D-printing "ink" for reproduction in outer space. As illustrated in the Introduction, 3D printing has been successfully tested in the ISS, and multiple reuse of printed materials is also technically feasible. Danielle Wood from MIT has proposed a new way of post-mission disposal. Her idea is that after their end-of-mission, satellites can be melted into 3D printer stock material to regenerate their successors.[109] This raises a question as to whether these newly printed satellites shall be regarded as "space object" under the UN space treaties.

When regulating ownership issues, the Outer Space Treaty uses the wording "objects launched into outer space, including objects landed or *constructed on a celestial body*, and of their component parts" (italics added).[110] It implies that an object that was first launched and then installed on the celestial body should still be regarded as an object launched.[111] As commented by Skopowska, this means "once launched, an object would never lose its quality of being an object launched".[112] It follows that the newly printed satellites, as they are made from Earth-launched materials, should be regarded as "space object". Moreover, a restrictive interpretation to exclude these newly printed objects from the scope of "space object" will result in legal lacunae, which is undesirable for peaceful and orderly operation of space activities. Therefore, a more inclusive approach should be adopted.

In this regard, Skopowska proposes to interpret the term "launch" so broadly as to include "all the possible ways of 'conveying' an object in the outer space, such as placing, installing or even producing an object in the outer space or on a celestial body".[113] According to this interpretation, newly printed objects fall within the scope of "space object" and their launching State is "the State at whose facilities an object

[107] Liability Convention, Article XI(2).

[108] UN Doc. A/AC.105/C.2/2013/CRP.6, Information on the activities of international intergovernmental and non-governmental organizations relating to space law (March 26, 2013).

[109] Meghan Bartels, "Changing How We Build Satellites Could Do More Than Reduce Space Junk" (January 24, 2019), https://www.space.com/43098-satellite-design-engineering-equity-justice.html.

[110] Outer Space Treaty, Article VIII.

[111] Laura Skopowska, "Is an object built in the outer space a 'space object' under the Liability Convention?" Master thesis, University of Luxembourg, 2017, 28.

[112] Ibid.

[113] Ibid.

was manufactured".[114] Meanwhile, the State which orders the 3D printing is also a launching State as it "procures" the launching. Notedly, whatever "launch" means, it can be inferred from the wordings "launched into outer space" and "return to the Earth"[115] that space objects are made from terrestrial materials.

It is interesting to refer to a report released by Joint Air Power Competence Centre which defines "space object" as "any man-made object in outer space".[116] In a similar vein, a joint proposal put forward by Argentina, Belgium and France in the drafting phase of the Liability Convention defines "space object" as "any object made or intended for space activities."[117] Reference can be made to the updated draft of the *Treaty on the Prevention of the Placement of Weapons in Outer Space, the Threat or Use of Force Against Outer Space Objects* (PPWT) as proposed jointly by China and Russia in 2014. Instead of referring to "space object", the 2014 PPWT uses "outer space object" which is defined as "any device placed in outer space and designed for operating therein".[118] It further specifies that "a device is considered as "placed in outer space" when it orbits the Earth at least once, or follows a section of such an orbit before leaving this orbit, or is placed at any location in outer space or on any celestial bodies other than the Earth".[119] These definitions emphasize the spatial location or the spatial purpose of space objects rather than how they enter outer space. Therefore, it can be argued what matters is the fact that an object originating from Earth is now placed into outer space. The way in which this result is achieved, whether by launching, placing, or manufacturing, is inconsequential.

Also, in the future, space objects may be manufactured from recycled materials together with extraterrestrial materials, or even entirely from the latter. These objects are not (entirely) made from Earth-launched materials, and therefore cannot be categorized as "space object". Reference in this regard can be made to the *Building Blocks for the Development of an International Framework on Space Resource Activities* (Building Blocks) adopted by The Hague International Space Resources Governance

[114] Ibid., 43.

[115] See, e.g., Outer Space Treaty, Article VIII.

[116] Joint Air Power Competence Centre, "Command and Control of a Multinational Space Surveillance and Tracking Network", by Lt Col Andrea Console (ITA AF), June 2019, 6. Notedly, the report states that "one can consider the following proposed definitions valid only for the purposes of this study, and thus devoid of any legal or political implications".

[117] UN Doc. PUOS/C.2/70/WG.I/CRP.16, Argentina, Belgium, France: Working Paper, in: UN Doc. A/ AC.105/85, Report of the Legal Sub-Committee on the Work of Its Ninth Session (8 June–3 July 1970), Annex 1, Proposals and Other Documents Relating to Liability For Damage Caused by the Launching of Objects into Outer Space (Agenda Item 2), 3 July 1970, 16. Citing from Smith and Kerrest, "Article I (Definitions) LIAB", *supra* note 51, 110.

[118] 2014 PPWT, Article I(a).

[119] 2014 PPWT, Article I(c).

Working Group (The Hague Working Group) on November 12, 2019.[120] The Building Blocks defines "space object" as "an object launched into outer space from Earth, including component parts thereof as well as its launch vehicle and parts thereof".[121] It uses another term "space-made product" to denote "a product made in outer space wholly or partially from space resources".[122] Therefore, for 3D printed objects, the origin of their 3D printing filament is of relevance to their legal status. More specifically, "space object" only covers objects made entirely from terrestrial materials, and a parallel regime should be established to regulate those made from extraterrestrial materials.

3.6 Conclusion

Apart from the overall benefits from scientific, social, and economic perspectives, over sixty years of space activities have also created a thorny problem—space debris. These defunct objects floating in space pose a threat to satellites and manned spacecraft. They can also self-generate in number through collisions and explosions. To preserve outer space as a sustainable environment for current and future generations, the space debris problem needs to be solved as soon as practicable. While the current efforts of the international community focus on space debris mitigation which aims to limit the creation of new debris, space debris remediation is also needed for the removal of existing debris in the future.

The reutilization of non-functional objects in space may also be an option to solve the debris problem. On-orbit servicing can resurrect these objects through refueling, repairing, upgrading, and re-orbiting. Furthermore, most of these objects are made from valuable materials and their recycling can save the effort in launching these materials from Earth. Apart from turning waste into treasure, space recycling also helps to reduce the number of space debris. Space actors and practitioners, both governmental and non-governmental, have put forward various proposals on space recycling projects, some of which have already been initiated.

Though the UN space treaties do not directly address space recycling, provisions contained therein are general enough to provide a framework for its regulation. These treaties are of direct relevance to governmental operators as their conduct shall be considered an act of their States. Meanwhile, they are also relevant to non-governmental operators as their responsible States should authorize and continually supervise their activities and assure that these activities are carried out in conformity with international law. Therefore, the coherence between international law and national space legislation can be expected.

[120] International Institute of Air and Space Law, "The Hague International Space Resources Governance Working Group", https://www.universiteitleiden.nl/en/law/institute-of-public-law/institute-of-air-space-law/the-hague-space-resources-governance-working-group.

[121] Building Blocks, Article 2.4.

[122] Building Blocks, Article 2.5.

A concept central to the UN space treaties is "space object", which "includes its component parts as well as its launch vehicle and parts thereof". Through literal, contextual, and purposive interpretations, space objects are those man-made objects launched into outer space, whether functional or not, which are physical and tangible. Accordingly, space recycling targets fall into this definition. Meanwhile, these targets, if non-functional, are also categorized as "space debris" under the UNCOPUOS SDM Guidelines. As the guidelines is of voluntary and non-legally binding nature, this categorization does not affect the legal status of space objects as accorded under the UN space treaties.

According to the Outer Space Treaty, the State of registry of a space object shall retain jurisdiction and control over such object. Therefore, space scavengers may only recycle a space object when their responsible State retains jurisdiction and control over such object or the State of registry so allows. Recall that only the launching States of a space object can register such object, the transfer of registration can only be made amongst these States. This poses a legal challenge when the responsible State of the space scavenger is not a launching State of the recycling target. One solution is the conclusion of an agreement between the responsible State and the launching States to arrange jurisdictional issues.

Meanwhile, since the ownership of space objects is not affected by their location, space scavengers should arrange proprietary issues with owners of recycling targets. The transfer of ownership is normally subject to the authorization of the responsible States of both space scavengers and owners. Due to the strategic importance of space objects, export control rules may also play a role in the transaction.

The Liability Convention elaborates on the liability rules contained in the Outer Space Treaty, which stipulates absolute liability for damage caused by a space object on the surface of the Earth or to aircraft in flight. Fault liability applies to damage caused by a space object in outer space. Whether the (non-)compliance of international soft law is pertinent to the determination of fault remains controversial. This article argues that those non-legally binding guidelines and standards should also be considered by judges to determine fault as this approach can to a greater extent encourage responsible behaviors in outer space. To reduce the risk of being held at fault, it would be in the interest of space scavengers to assure their adherence to the international guidelines and standards. They may also refer to the Long Term Sustainability Guidelines as a defense against fault if damage occurs.

The advancement of 3D printing technologies creates a new possibility for space recycling, as recycled materials may be used as 3D printing filament to manufacture new objects. These objects are not directly launched from Earth but are made from Earth-launched materials. Excluding these objects from the scope of UN space treaties will lead to undesirable results, and therefore the term "launch" should be interpreted broadly to include, as proposed by Skopowska, "all the possible ways of 'conveying' an object in the outer space".[123] In this regard, reference can be made to several international proposals and instruments which use the terms "made" and "placed" instead of "launched".

[123] Skopowska, *supra* note 111, 43–44.

With the development of space resource utilization, in the future news objects may be made in space from a mixture of recycled materials and extraterrestrial materials. Regarding their legal status, as the wordings "launched into outer space" and "return to the Earth" used in the UN space treaties indicate the terrestrial origin of "space object", those objects made from extraterrestrial materials do not fall within this scope. Reference can be made to the Building Blocks adopted by The Hague Working Group, which distinguishes "space object" from "space-made product". While the former bears a similar meaning defined under the UN space treaties, the latter refers to "product made in outer space wholly or partially from space resources".[124]

The UNCOPUOS has adopted various sets of principles to address specific categories of space activities such as broadcasting and remote sensing.[125] Meanwhile, some national laws even take one step ahead of space technology development, such as the U.S. Space Resource Exploration and Utilization Act of 2015 which prescribes proprietary rights of U.S. citizens to any asteroid resource or space resource obtained.[126] Recall that space recycling contributes to the protection of outer space environment which is generally regarded as a global commons and that it normally involves the transfer of jurisdiction between two or more nations, an international approach is preferable. A potential path forward for its governance is the establishment of a non-legally binding instrument under the auspices of the UN, most ideally through the adoption of a UN General Assembly resolution. This instrument may obtain binding force through implementation at the national level, and it may also evolve into customary international law if consistent and uniform *opinio juris* and state practice can be observed.

Zhuang Tian was born and raised in Guangzhou, China. He is currently a Ph.D. candidate at Leiden University, the Netherlands, where he has also obtained his LL.M. degree. His research interests lie in space law, especially the legal aspects of space debris mitigation and remediation.

Yangyang Cui was born and raised in Ningbo, China. She is currently an LL.M. candidate at Shanghai Jiao Tong University. Her research interests lie in international public law and international trade law.

[124]Building Blocks, Article 2.4 and 2.5.

[125]UNOOSA, "Space Law Treaties and Principles", https://www.unoosa.org/oosa/en/ourwork/spacelaw/treaties.html.

[126]51 U.S. Code § 51303.

Chapter 4
On-Orbit Servicing: Security and Legal Aspects

John Tziouras

Abstract On-orbit servicing capabilities seem to be a "game changer" for Space 2.0. Recent advances in the field of on-orbit servicing (OOS) of defunct satellites and other space infrastructure have increased the need for the development of a comprehensive legal regime for the new Space Industry at an international level. The main challenges for OOS lie in several different realms, both legal and technical, but with perhaps a common issue: space security. The dual-use nature of OOS has given rise to concerns that these technologies could be used for military purposes in space. Reducing the perceived and actual risks is necessary for investors to move in. One probable solution would be for practitioners to develop a bottom-up lawmaking approach using existing or new *fora*. When there are stalemates and traditional top-down approaches fail, legal history has shown that practice-based rules become law.

4.1 Introduction

In December 2019, US company Northrop Grumman launched the world's first commercial robotic satellite servicing mission, pioneering what is projected to become a multi-billion dollar market over the next few years.[1] Satellites today are built to be disposable. The capability to refuel, repair and upgrade them robotically could transform commercial, civil and military space activities over the coming decade.

[1]Oxford Analytica Daily Brief, 'Space Servicing Industry Could Destabilize Geopolitics' (Daily-brief.oxan.com, 12 December 2019) https://dailybrief.oxan.com/Analysis/DB249391/Space-servicing-industry-could-destabilise-geopolitics, accessed 2 March 2020.

J. Tziouras (✉)
Faculty of Law, Aristotle University of Thessaloniki, Thessaloniki, Greece
e-mail: tziouras@law.auth.gr

© The Editor(s) (if applicable) and The Author(s), under exclusive license to Springer Nature Switzerland AG 2020

A. Froehlich (ed.), *On-Orbit Servicing: Next Generation of Space Activities*, Studies in Space Policy 26, https://doi.org/10.1007/978-3-030-51559-1_4

However, On-Orbit Servicing (OOS) is not a brand-new idea. Although orbit rendezvous missions can be traced back to the Apollo era, traditional human OOS, using Extravehicular Activity, has been conducted since 1980s.[2] One of the most spectacular missions was the first servicing mission that aimed at repairing an optic flaw in the primary mirror of Hubble Space Telescope resulting in blurry observations.[3] Since then, OOS activities have been performed on systems such as the Hubble Space Telescope and the International Space Station.[4]

Today, OOS can expand the possibilities offering new prospects for the space economy, while ensuring long-term sustainability of outer space exploration and use.[5] Most importantly, such activities are feasible by not involving humans. Though servicing promises to ensure space sustainability, at the same time it creates greater potential for space security. While OOS are not *stricto sensu* a military space activity, the core capability of maneuver, rendezvous and manipulation is of a dual-use nature.[6] For example, if a non-cooperative satellite can be serviced or deorbited, then it will be much easier to perform the same operation with a cooperative satellite.[7] These development efforts, including Active Debris Removal (ADR) actions, demonstrate an important concern many have regarding these technologies.[8]

In an effort to shed some light on a number of aspects related to international space law, space security and OOS the emphasis on this chapter will be on the legal concepts of dual-use nature, liability matters and security aspects of OOS.

---

[2]International Space University, 'DOCTOR: Developing On-Orbit Servicing Concepts, Technology Options, and Roadmap' (2007), Final Report, Summer Session Program.

[3]C. Joppin, *On-Orbit Servicing for Satellite Upgrades* (Massachusetts Institute of Technology 2004) 28.

[4]D. Hastings, E.S. Lamassoure, A.L. Weigel, J.H. Saleh, 'Policy Enablers for the Development of a Space-Based Infrastructure', in W.A.H. Thissen and P.M. Herder (eds) *Critical Infrastructures: State of the Art in Research and Application* (Springer Science + Business Media 2003) 124.

[5]M. Frigoli, 'Between Active Debris Removal and Space-Based Weapons: A Comprehensive Legal Approach' in Annette Froehlich (ed) *Space Security and Legal Aspects of Active Debris Removal* (Springer Nature Switzerland AG 2019) 52.

[6]P.J. Blount, 'On-Orbit Servicing and Active Debris Removal: Legal Aspects' in A.N. Pecujlic and M. Tungoli, *Promoting Productive Cooperation Between Space Lawyers and Engineers* (IGI Global 2019) 180.

[7]J. Alver, A. Garza, C. May, An Analysis of the Potential Misuse of Active Debris Removal, On-Orbit Servicing, and Rendezvous & Proximity Operations Technologies, Elliot School of International Affairs, The George Washington University (swfound.org, 6 May 2019) https://swfound.org/media/206800/misuse_commercial_adr_oos_jul2019.pdf, accessed 2 March 2020.

[8]B.G. Chow, 'Space Arms Control: A Hybrid Approach' (2018) 12 Strategic Studies Quarterly, 107–132.

4.2 Space Security and OOS: Setting the Appropriate Security Study Context

As Moltz states,[9] during the Cold War, the political relations of the two superpowers dominated space security issues. In that context, space security was focused on military and environmental aspects of accessing space. Since non-governmental entities have made their presence in outer space in the last decades and new space commercial and resources activities have been taking place, it was necessary to broaden the security concept.

A broader definition should encompass both natural and human threats, addressing the issues of environmental and military security risks, covering different dimensions which are interconnected. According to Mayence space security means at the same time:

- outer space for security: the use of space systems for security and defense purposes,
- security in outer space: how to protect space assets and systems against natural and/or human threats or risks and to ensure a sustainable development of space activities, and
- security from outer space: how to protect human life and earth's environment against natural threats and risks from outer space.[10]

"*The advantage of expanding conceptions of space security is that it moves certain issues significantly higher up the political agenda*",[11] says Sheehan. Nevertheless, he observes it is also important to note that the move to broaden understandings of security carries certain risks. "*If more and more issues are brought into the realm of "security," there is logically a point at which the term is so all encompassing that in practice it means very little and becomes simply another word for "dangers" or "risks," and eventually the concept would have no coherent meaning and therefore no value as a guide to policy.*"[12]

For that reason, it is very important to limit the context of security issues,[13] otherwise expanding the term "security" to new areas such as the OOS "threats" or dangers, will end up being subsumed within a military context which is inappropriate, making the problem harder to deal with.

---

[9] J.C. Moltz, *The Politics of Space Security: Strategic Restraints and the Pursuit of National Interests* (Stanford University Press 2011) 11.

[10] J.F. Mayence, 'Space Security: Transatlantic Approach to Space Governance' in J. Robinson, M.P. Schaefer, K-U Schrogl, F. von Der Dunk (eds) *Prospects for Transparency and Confidence-Building Measures in Space*, European Space Policy Institute Report 27, (ESPI 2010) 35.

[11] M. Sheehan, 'Defining Space Security' in K-U Schrogl, P.L. Hays, J. Robinson, D. Moura, Ch. Giannopapa (eds) *Handbook of Space Security* (Springer Science + Business Media New York 2015) 19.

[12] Ibid.

[13] Ibid., 20.

4.2.1 Challenges for Space Security Associated with OOS

In order to understand the concern arising for national security from OOS technologies (under development or potential) it is necessary to define the general requirements for operations of OOS systems. All OOS and ADR systems will require Rendezvous and Proximity Operations (RPO) technologies in order to operate.[14]

RPO capabilities require for example camera and/or radar technologies, the ability to process data in near real-time, as well as propulsive capabilities to maneuver along very defined, near-target trajectories. In light of the above, there are two main approaches that so far have been developed to address the security issues of OOS.

First, according to USAF Doctrine on Counterspace Operations, "*RPO are specific processes where two resident space objects are intentionally brought close together.*"[15] Servicing of space assets requires the capabilities of RPOs. As a result, OOS capabilities enable inspection, repair, replacement, and/or upgrade of spacecraft subsystems components and replenishment of spacecraft consumables (e.g. fuels, fluids, cryogens).[16]

The second approach takes its momentum from recent strategic context, that OOS technologies could possibly lead to implied space weapons (Anti-satellite-ASAT). The primary concerns for RPO enabled systems arise from other activities that may be performed while in close proximity to other systems (i.e. jamming, grappling technologies, harpoons shots) with the intention of disabling critical components or the entire systems; satellites. However, as OOS operation would take several hours until mission completion[17] (even if the servicing system is pre-installed on space), it is obvious that these operations are not rapid events. As such, these requirements necessitate longer timelines, giving the "target" system operator the time to determine that another system is potentially approaching them.

Furthermore, since there are more developed "hazard free" ASAT technologies (i.e. Directed Energy Weapons- DEW), both ADR and OOS should not be considered as such for a couple of reasons (i.e. strategic value). Nevertheless, the risks or challenges these technologies pose for military or civilian satellites should not be underestimated completely. Frigoli,[18] at his significant research, theorise that as OOS

[14] J. Alver, A. Garza, C. May, An Analysis of the Potential Misuse of Active Debris Removal, On-Orbit Servicing, and Rendezvous & Proximity Operations Technologies, Elliot School of International Affairs, The George Washington University (swfound.org, 6 May 2019) https://swfound.org/media/206800/misuse_commercial_adr_oos_jul2019.pdf, accessed 2 March 2020.

[15] USAF Doctrine on Counterspace Operations, Procedures for Space Rendezvous and Proximity Operations, JP 3-14, Annex 3-14 (10 April 2018).

[16] Ibid.

[17] P. Colmenarejo, G. Binet, L. Strippoli, T.V. Peters, M. Graziano (eds) GNC Aspects for Active Debris Removal, Proceedings of the EuroGNC 2013, 2nd CEAS Specialist Conference on Guidance, Navigation & Control, Delft University of Technology, The Netherlands (aerospace-europe.eu, 10–12 April 2013) https://aerospace-europe.eu/media/books/delft-0084.pdf, accessed 2 March 2020.

[18] M. Frigoli, 'Between Active Debris Removal and Space-Based Weapons: A Comprehensive Legal Approach' in Annette Froehlich (ed) *Space Security and Legal Aspects of Active Debris Removal* (Springer Nature Switzerland AG 2019).

and ADR share the same technologies (but different techniques), there are concerns regarding distinction between the use of these technologies and the weaponization of outer space.

However, as Blount states, if OOS were used for ASAT purposes they would violate Art. 2 par. 4 of the UN Charter and "*while the space law does not prohibit OOS technologies de facto, it does prohibit certain use of that technology that impinge on the rights of other states by interfering with those states' space activities.*"[19]

Before supporting the view that OOS has to be conceived as a threat and not as another security issue, one should follow a methodological approach and examine some criteria about this argument.

4.2.2 Criteria on (De)Securitizing OOS

It is widely accepted that the space security agenda is being dominated by the so-called "widening vs narrowing" debate. Bowen shed light on this sensitive debate in his research[20] by examining the two main approaches: The arguments between the traditionalists and between those who are in favor of widening the agenda of (space) security. As Bowen[21] argues, "*framing a problem as a security issue usually means that a threat could escalate into an issue of survival and it may be difficult to escalate the issue of debris to one of survival.*"

Those in favor of widening security issues, such as Moltz, who enriches the space security debate with the importance of the space environment, argues that "*moving forward with a weapon-based approach toward security is more difficult to reverse that an approach based on international cooperation.*"[22]

The difference in approach lies at the referent object. For strategists the referent object is the state security while the referent object for others (i.e. Moltz) is the space environment itself. Still, as Moltz suggests, "*broader definitions of security appear to be more appropriate to space, given its mixed activities*"[23] something that a state-centric narrow perspective of security ignores.

While taking under careful consideration the differences between the Welsh School of international security studies and the Copenhagen School led by Buzan, Weaver and others, which addresses the shortcomings in realism approaches, the

---

[19]P.J. Blount, 'On-Orbit Servicing and Active Debris Removal: Legal Aspects' in A.N. Pecujlic and M. Tungoli, *Promoting Productive Cooperation Between Space Lawyers and Engineers* (IGI Global 2019) 182.

[20]B.E. Bowen, 'Cascading Crises: Orbital Debris and the Widening of Space Security' (2014) 12 Astropolitics: The International Journal of Space Politics & Policy, 46–68.

[21]Ibid., 52.

[22]J.C. Moltz, *The Politics of Space Security: Strategic Restraints and the Pursuit of National Interests* (2nd edition Stanford University Press 2011) 351.

[23]Ibid.

need for specializing in the space security agenda should not be rejected. Nonetheless, one should do so by setting under careful methodological control the issue that has to be securitized, in order to avoid seeing ghosts where they do not exist.

Hesse and Hornung[24] and other scholars[25] could probably help framing the OOS as an advanced security issue by theorizing space as a Critical Infrastructure (CI). The main question that arises here is if OOS could be considered such a threat to CI, capable of triggering security concerns due to the fact that dual-use of OOS is being conceived as a threat to active and functional satellites.

At the following section an effort will be made to demonstrate that using a broadened concept of security involves unnecessary risks that could diminish the significance of OOS and their ability to contribute to the solution of orbital debris and the capability of positively transforming the operational landscape of space activities.

4.2.3 Dual-Use of OOS as a Potential Threat but not as Another Security Issue

Despite the various ADR and OOS proposals, these operations require physical interaction between space objects. For this reason, there are serious concerns about the intention, these systems to be used for nefarious purposes leading to the weaponization of space. While the main purpose of such technologies is not military, however, as Chow mentioned "*a spacecraft that can perform OOS, can also destroy a satellite.*"[26]

OOS definitely cannot be considered as a space weapon. Should, though, be considered as a terrestrial or space capability that is capable of producing weapons effects in space? The analysis should focus on the potential misuse of OOS and the level of nonreversible effects such actions can cause to other systems. According to US Defense Intelligence Agency[27] DEW, cyber threats and orbital threats can cause both temporary or permanent effects. Some authors[28] consider the capabilities of unmanned proximity operations as a condition in order to theoritise that possible misuse can lead to an ASAT use.

---

[24]M. Hesse and M. Hornung, 'Space as a Critical Infrastructure' in K-U Schrogl, P.L. Hays, J. Robinson, D. Moura, Ch. Giannopapa (eds) *Handbook of Space Security* (Springer Science + Business Media New York 2015).

[25]L. Mureşan and A. Georgescu, 'The Road to Resilience in 2050: Critical Infrastructure and Space Security' (2015) 160 The Royal United Services Institute Journal.

[26]B.G. Chow, 'Space Arms Control: A Hybrid Approach' (2018) 12 Strategic Studies Quarterly, 108.

[27]Defense Intelligence Agency, 'Challenges to Security in Space' (Dia.mil, January 2019) https://www.dia.mil/Portals/27/Documents/News/Military%20Power%20Publications/Space_Threat_V14_020119_sm.pdf, accessed 3 March 2020, 36.

[28]B.G. Chow, 'Space Arms Control: A Hybrid Approach' (2018) 12 Strategic Studies Quarterly, 107–132.

By adopting the above-mentioned approach, one runs the risk to conceive every satellite with maneuvering capabilities as a potential hostile orbital system. Moreover, all ADR and OOS systems since their conceptual phase are at risk of being conceived as potential ASAT weapons. For these reasons it is crucial to distinguish between definitions, conceptions, interpretations and uses.

4.3 Dealing with the Perception of "Others" and OOS

As Wolf states,[29] the seeming contradiction over peaceful uses of outer space originates from the fact that conventional space law never precisely defines the term "peaceful". "*With ambiguous definitions subject to various interpretations, certain activities that one would not normally consider peaceful have been pursued*"[30] as non-aggressive.

Yet, many activities are linked to the beliefs of others. The U.S. Space Shuttle, which was an early entrant into OOS technology, is rarely considered today as having had an anti-satellite capability, but during the first years, the Soviet Union had objections to the shuttle's ability to rendezvous with satellites and drag them inside.[31]

From our point of view, it seems important, during this phase of OOS development, for stakeholders to work on Transparency and Confidence-Building Measures (TCBMs). To achieve this aim, as Frigoli explains, approaching the issue from the point of view of ADR,[32] a key-role should be played by the mandatory international consultation in case of a potentially harmful interference with activities in the peaceful exploration and use of outer space, which could deny access to outer space. Still, as some OOS development projects are funded by military organizations (i.e. DARPA) specific questions may arise as to whether this leads to weaponization of space and if so what laws there are to limit such weaponization.[33]

In light of all these, the following legal analysis will attempt to explain how this (not all) new category of space mission may fit or not to the current legal framework.

---

[29] J.M. Wolf, 'Peaceful Uses of Outer Space has Permitted its Militarization-Does it Also Mean its Weaponization?' (2003) 1 Disarmament Forum 5.

[30] Ibid.

[31] L. Grego, 'A History of Anti-Satellite Weapons' Union of Concerned Scientists (ucsusa.org January 2012) https://www.ucsusa.org/sites/default/files/2019-09/a-history-of-ASAT-programs_lo-res.pdf. Accessed 2 March 2020.

[32] M. Frigoli, 'Between Active Debris Removal and Space-Based Weapons: A Comprehensive Legal Approach' in Annette Froehlich (ed) *Space Security and Legal Aspects of Active Debris Removal* (Springer Nature Switzerland AG 2019).

[33] P.J. Blount, 'On-Orbit Servicing and Active Debris Removal: Legal Aspects' in A.N. Pecujlic and M. Tungoli, *Promoting Productive Cooperation Between Space Lawyers and Engineers* (IGI Global 2019).

4.4 Legal Aspects Concerning OOS Activities

While OOS and ADR technologies have great similarities, as both contribute to the continued sustainability of the space environment, one of the main differences between them is that the main purpose of OOS is the cost-effective alternative to traditional methods of space systems maintenance.[34] The primary benefit of ADR will be realized only in the long term as it will reduce the possibility of unintentional accidents. However, OOS offers short-term benefits by way of mission life extension, thus extending the capabilities for both governments and private sector on their space activities.[35]

Such differences on the initial purpose of each technology derive from the different legal aspects on the one hand and on the different legal approaches on the other. Currently no national or international text explicitly regulates OOS, but rather the space law regime covers a great part of these technologies, as it will be explained below. Blount in his comprehensive legal analysis[36] examines whether OOS fits into the current legal framework. It must be agreed then that "*these technologies do not constitute a space weapon per se*" and any ASAT use of OOS violates the rights of other states under the Art. 2 par. 4 of the UN Charter. However, as he states, this does not mean that there are no serious concerns about the possible perception of these technologies as ASAT.

The Treaty on Principles Governing the Activities of States in the Exploration and Use of Outer Space, Including the Moon and Other Celestial Bodies (OST) does not clearly address ASAT or the placement of conventional weapons in space.[37] As a result, it is necessary to rely on the responsibility provisions for activities in outer space laid down by OST and general international law, as well as on principles of the prohibition of harmful contamination of outer space and harmful interference of outer space activities which derives from Art. I and IX of OST.[38] Also, under general international law both States and International Organizations bear international responsibility for their activities. Responsibility for internationally unlawful acts must be differentiated from the liability regime set out in Art. VII of OST

[34]Y. Porat et al., 'The On Orbit Servicing Answer to Safety and Sustainability for Future Space Activities', 67th International Astronautical Congress, Guadalajara, Mexico, 26–30 September 2016.

[35]UNCOPUOS (Scientific and Technical Subcommittee), 'Active Debris Removal—An Essential Mechanism for Ensuring the Safety and Sustainability of Outer Space' (27 January 2012) UN Doc A/AC.105/C.1/2012/CRP.16, 28.

[36]P.J. Blount, 'On-Orbit Servicing and Active Debris Removal: Legal Aspects' in A.N. Pecujlic and M. Tungoli, *Promoting Productive Cooperation Between Space Lawyers and Engineers* (IGI Global 2019).

[37]J.M. Wolf, 'Peaceful Uses of Outer Space has Permitted its Militarization-Does it Also Mean its Weaponization?' (2003) 1 Disarmament Forum 5.

[38]M. Frigoli, 'Between Active Debris Removal and Space-Based Weapons: A Comprehensive Legal Approach' in Annette Froehlich (ed) *Space Security and Legal Aspects of Active Debris Removal* (Springer Nature Switzerland AG 2019).

and the Convention on International Liability for Damage Caused by Space Objects (Liability Convention) which will discuss next.

If this problem is approached through the traditional arms controls lenses, it will result in ending up in the same stalemate with ASAT and Prevention of an Arms Race in Outer Space (PAROS) negotiations. The only way to move forward is to distinguish between the uses of OOS. The fact that there is no effective arms control regime or a treaty banning ASAT does not mean that OOS activities should be uncontrollable. As Chow proposes,[39] OOS can be controlled by prohibiting certain activities (i.e. not being simultaneously placed too close to and threatening another country's satellite etc.).

While the concerns about OOS potential "destabilizing" uses have to be regulated, one should also focus on other legal aspects that have to be examined, such as international liability and licensing issues, insurance aspects, intellectual property rights and standardization (these issues will not be addressed here) etc. As OOS activities seem to be carried out by commercial, non-governmental entities, it is important to address some issues associated with the commercial deployment of OOS.

4.4.1 Liability, Responsibility and Licensing Issues

When damage is caused by a space object in outer space, Article VII of OST alongside with the Liability Convention provides a mechanism for compensation for the injured state. Liability Convention has set up a regime of joint and several liability resting upon the state parties to other state parties for damage caused by a space object for which they each qualify as a launching state. The liability in respect of damages caused to third parties is divided into two parts: a fault-based liability for damage to other spacecraft occurring in space and non-fault-based liability and an absolute liability for damage caused by its space object, though in the former case only to the extent of its fault.[40] One of the aspects here concerns the meaning of the term "fault" under Art. III of the Liability Convention. Some authors point out that it is unclear whether the treaty refers to fault in the context of a fault liability regime or fault as understood within the regime of state responsibility for wrongful acts. How this could affect OOS? Dennerley,[41] in order to differentiate between the two methods of interpretation concerning the term "fault" (under responsibility regime and liability

[39] B.G. Chow, 'Space Arms Control: A Hybrid Approach' (2018) 12 Strategic Studies Quarterly, 107–132.

[40] C. Gaubert, 'Insurance in the Context of Space Activities' in F. von der Dunk and F. Tronchetti, *Handbook of Space Law*, (Edward Elgar Publishing 2015) 910.

[41] J.A. Dennerley, 'State Liability for Space Object Collisions: The Proper Interpretation of 'Fault' for the Purposes of International Space Law' (2018) 29 European Journal of International Law 281–301.

regime), suggests that this may have repercussions on compensation and reparation in case of space object collisions.

As there is no insurance obligation by either the OST of Liability Convention and remedies provided by Liability Convention are available solely between states, there are some legal issues to be addressed, especially when OOS activities are undertaken by private actors. The OST gives States the duty to directly oversee non-governmental actors. However, as there are significant national laws concerning licensing and insurance regulations, the gaps in authorization and continuing supervision at the domestic level create a legal vacuum in which two forces work to cancel each other out. As Blount states precisely, "*without regulatory certainty investors are unwilling to invest and without investment from industry, regulators are not pushed to develop rules.*"[42]

4.4.2 Space Insurance

The space insurance industry is a high risk area which is not regulated on an international basis.[43] As, Gaubert comprehensively explains, "*there are two types of space insurance: one covering first-party property insurance and the other, dealing with third-party liability insurance, addressing third-party liability of a launching agency or satellite operator/owner whose launcher, satellite or part thereof is considered accountable for damages caused to third parties during the space operation*",[44] as it was mentioned above.

Contrary to third-party liability insurance, no insurance obligations resting upon the space operator exist under international or national law. The decision to insure orbital property rests entirely with the operator or their clients. The insurance contract has been exclusively within the realm of the private sector, playing the role of mollifying investors in a space program that their investments would be safe and covered by insurance in the event of damage or launch failure. When talking about space property damage insurance, three periods of risks are involved, being the pre-launch, launch and in-orbit period, should be identified.

Insurance of space activities and spacecraft amply reflects the significance of risk management. In the present context of OOS, risk management becomes critical for both the insured and the investor in relation to all types of insurance (especially third party liability), as OOS operations entail greater risk than routine satellite orbit maintenance. As Reesman points out, there are different aspects concerning OOS: "*The servicing vehicle may carry liability insurance, which would include*

[42]P.J. Blount, 'On-Orbit Servicing and Active Debris Removal: Legal Aspects' in A.N. Pecujlic and M. Tungoli, *Promoting Productive Cooperation Between Space Lawyers and Engineers* (IGI Global 2019).

[43]R. Abeyratne, *Space Security Law* (Springer-Verlag Berlin Heidelberg 2011) 67.

[44]C. Gaubert, 'Insurance in the Context of Space Activities' in F. von der Dunk and F. Tronchetti, *Handbook of Space Law*, (Edward Elgar Publishing 2015) 910.

launch and performance capabilities covering loss, damage or failure, while the satellite to be serviced may or may not be insured, but a contract with the servicer would outline expectations. Additionally, third-party liability insurance would cover damages imposed on an asset not involved in the servicing agreement."[45]

Furthermore, according to what has already been addressed above relative to risk management, OOS could be used as a risk-mitigation tactic. Servicing or repair of a satellite could help to avoid catastrophic loss and could prevent the need for an insurance payout.[46]

4.5 What Is Next? A Bottom-Up Lawmaking Approach

In the age of privatization of space activities new non-governmental actors made their presence opening up, or at least planning to open up, new markets in the new Space Industry. This emerging trend of private enterprises is characterized by creative "out-of-the-box" thinking, leading to revolutionary ideas and concepts. While State "space secrets" are still well-kept, the new Space Industry wants to be what the older traditional space industry is not: a non-bureaucratic and entirely cost-effective and not top-down governmental way of doing things. In an era of coexistence, traditional space industry and New Space Industry turn out to have a symbiotic relationship.[47]

The same symbiotic relationship that exists between international space law and national space legislation. In this so-called "fourth phase" of space law (*Lato Sensu*) as von der Dunk explains, the changing role of law-making follows the new trend. So, today, "*space law should not be taken to refer only to those global treaties, resolutions and other legal or soft-law developments which principally originated from the bosom of COPUOS, or more precisely from the cooperation between most of the major spacefaring states.*"[48] In this case, bottom-up lawmaking,[49] as an alternative to the traditional top-down law making, do not feature state policymakers but rather the very practitioners-both public and private. This process has a great advantage concerning the previous experience that practitioners usually have. As a result, a more detailed,

---

[45]R. Reesman, 'Assurance Through Insurance and On-Orbit Servicing' Center for Space Policy and Strategy, The Aerospace Corporation, (Aerospace.org, May 2018) https://aerospace.org/sites/default/files/2018-05/OnOrbitServicing.pdf. accessed 28 February 2020.

[46]Ibid.

[47]E. Jonckheere, 'The Privatization of Outer Space and the Consequences for Space Law', Faculty of Law and Criminology Ghent University, (2018) | J.K. Levit, 'A Bottom-Up Approach to International Lawmaking: The Tale of Three Trade Finance Instruments' (2005) 30 The Yale Journal of International Law 129–132.

[48]F. von der Dunk, 'International Space Law' in F. von der Dunk and F. Tronchetti (eds), *Handbook of Space Law*, (Edward Elgar Publishing 2015).

[49]J.K. Levit, 'A Bottom-Up Approach to International Lawmaking: The Tale of Three Trade Finance Instruments' (2005) 30 The Yale Journal of International Law 129–132.

carefully drafting process begins which over time leads to the initially non-binding rules becoming law.

In the light of the above mentioned, space nowadays, is a privileged area for bottom-up lawmaking for a couple of reasons: First, UN space law was adopted at the time when states were the only actors in this field when space activities were carried out mainly for strategic purposes.[50] Today national legislation, state best practices or organization driven efforts (i.e. DARPA[51] about OOS), start framing a new legal landscape for commercial activities. Secondly, traditional arms control approaches, on Conference on Disarmament (CD), are still in a stalemate preventing any development in that field. While very useful tools were developed on international UN Committee forums/fora on the Peaceful Uses of Outer Space (COPUOS), the contribution of non-state actors, including institutions (i.e. European Space Policy Institute, Secure World Foundation), should be upgraded.

A possible way to develop regulatory mechanisms or legal instruments, capable of regulating OOS activities, is to involve all the relevant stakeholders—especially non-governmental actors—in an alternative, bottom-up, lawmaking approach based initially on informal, practice-based rules which lead to more formal legal instruments at national or international level. There are analogous success stories in the realm of international trade law.

Janet Koven Levit, in a very useful article mentioned that bottom-up lawmaking generates a specific, technical regulatory system. As she pointed out, "*transnational legal processes typically begin with a treaty or some other legal instrument that triggers a trickle-down process whereby domestic legal systems absorb international law. In contrast, bottom-up lawmaking features soft rules ascending to legal status.*"[52] Such processes start with a relatively small, homogenous lawmaking group which creates substantive rules, which are essentially organic norms emanating from the practices of the respective practitioners. While Levit studies the success of bottom-up lawmaking practice in the case of several international trade legal treaties, there seem to be various analogies with OOS.

In that sense, non-state actors, such as those mentioned above, as well as OOS practitioners (even from the public sector) could initiate a process, pursued outside the existing multilateral fora, bypassing at least stalemates on PAROS. Also, given the divergent purpose relative to ADR, OOS bottom-up lawmaking process initiative could be studied autonomously, as this is mainly this book's aim. While there is a recently beautiful work on ADR, OOS should be studied under its own theoretical, strategic, legal and political context.

---

[50] F. von der Dunk, 'International Space Law' in F. von der Dunk and F. Tronchetti (eds), *Handbook of Space Law*, (Edward Elgar Publishing 2015).

[51] D.A. Barnhart and R. Rughani, 'On-Orbit Servicing Ontology applied to Recommended Standards for Satellite in Earth Orbit' (2019) 70th International Astronautical Congress (IAC), IAC-19-D1.6.9, Washington D.C., International Astronautical Federation.

[52] J.K. Levit, 'A Bottom-Up Approach to International Lawmaking: The Tale of Three Trade Finance Instruments' (2005) 30 The Yale Journal of International Law 129–132.

4.6 Conclusion

At his research, de Waal Alberts[53] compares the existing analogies between aviation and space under a methodological approach using an Oxford Analytica Report. The author theorizes that the turning point of the aviation industry into a fully fledged commercial one was caused by the introduction of a unique technological innovation (jet engine). Could recent developments on SpaceX's reusable launch system or on OOS technologies as the condition upon which the growth of a new Space era will occur be accepted?

Undoubtedly, OOS capabilities seem to be a game changer. Even if the area of OOS will appear not to be the equivalent of a jet engine, on-orbit services may prove central to much of what happens in space in years to come. OOS technologies as Pelton[54] describes, may prove key to removing space junk from orbit, fixing anomalies or extending the service life of aging satellites, manufacturing and processing of materials in space, building solar power satellites, etc.

While these new technologies developed by governmental space agencies and organizations, as national security matters are dominant, the commercial character of these technologies are increasingly involving the private sector. Moreover, as neither national nor international (space) law directly addresses OOS capabilities, there is a need to develop special norms and law in order to tame uncertainty and foster investment and innovation.

Nonetheless, the dual-use nature of OOS technologies bears different practical and legal consequences which need to be further regulated. Specific to this discussion, the dual use capabilities of OOS raise questions related to the weaponization of space, and as a result OOS becomes part of the PAROS agenda ending up in the same stalemate. As Martinez summarizes, "*there does not seem to be consensus on the desirability of legally binding instruments banning the placement and use of space weapons.*"[55]

---

[53] A. de Waal Alberts, 'The Probable Contribution of the Post-2030 Space Industry to Global Economic Development' in A. Froehlich (ed), *Post 2030-Agenda and the Role of Space: The UN 2030 Goals and Their Further Evolution Beyond 2030 for Sustainable Development*, (Springer International Publishing 2018) 78.

[54] J.N. Pelton, 'On-Orbit Servicing, Active Debris Removal and Repurposing of Defunct Spacecraft' in J.N. Pelton (ed), *Space 2.0 Revolutionary Advances in the Space Industry* (Springer Nature Switzerland 2019).

[55] P. Martinez, 'Space Sustainability' in K-U Schrogl, P.L. Hays, J. Robinson, D. Moura, Ch. Giannopapa (eds) *Handbook of Space Security* (Springer Science + Business Media New York 2015) 271.

For this reason, the development of voluntary frameworks for promoting space sustainability (as this is the main purpose of OOS) provides some scope for making progress. By involving all the relevant stakeholders (especially non-governmental ones) in developing through a bottom-up lawmaking approach, national and international instruments, any traditional stalemates could be surpassed. At the very end, governments can choose to domesticate or not those non-binding (soft law) instruments, as they are accountable for their choices over their citizens.

John Tziouras is a Ph.D. candidate at Aristotle University of Thessaloniki, Faculty of Law (Greece). His research interest is in space security and space law. He obtained his LL.M. degree in International and European Legal Studies at Aristotle University of Thessaloniki. He was member of European Center of Space Law and now works as a special advisor for local authorities.

Chapter 5
On-Orbit Servicing: Space for Africa?

André Siebrits

Abstract On-orbit servicing (OOS) has the potential to revolutionise the space industry through the extension of satellite lifespans, in-situ repairs and upgrades, and active debris removal. The sector has also undergone its own historical evolution, of which the current first steps in robotic OOS is the latest phase. This is of great interest not only to developed space actors, but also to the developing space actors of the Global South, including Africa. In recent years, rapid advances have been made in this continent's space sector, with increasing numbers of African satellites serving social, political, and economic development goals. However, several expensive satellite failures or impairments have also hampered these efforts, and OOS can thus play a vital role for Africa. The establishment of the African Space Agency and the creation of the African Space Policy and Strategy also lay the foundation for the continent's potential future role as a supplier, not only a consumer, of OOS. However, significant challenges remain, and the instructive case of Edward Mukuka Nkoloso, a failed pioneer of the African space sector, is used to frame the chapter's approach to these issues.

> I see the Zambia of the future as a space-age Zambia, more advanced than Russia and America. ... If I had had my way Zambia would have been born with the blast of [the] academy's rocket being launched into space. ... I feel the Zambian Government should help now if we are to become Controllers of the Seventh Heaven of Interstellar space. ... The capital of the new scientific Zambia must look beautiful. People from afar must not see a slum as the capital of the world's greatest scientific state. Zambians are inferior to no men in science technology. My space plan will surely be carried out.
>
> Edward Mukuka Nkoloso[1]

[1] Edward Makuka Nkoloso, "We're going to Mars! With A Spacegirl, Two Cats and a Missionary", *Lusaka Times*, 28 January 2011, https://www.lusakatimes.com/2011/01/28/space-program/ (all websites cited in this chapter were last accessed and verified on 31 March 2020).

A. Siebrits (✉)
European Space Policy Institute, Vienna, Austria
e-mail: andre.siebrits@southernspacestudies.com; SBRAND003@uct.ac.za

Department of Political Studies, University of Cape Town, Rondebosch, South Africa

© The Editor(s) (if applicable) and The Author(s), under exclusive license to Springer Nature Switzerland AG 2020

A. Froehlich (ed.), *On-Orbit Servicing: Next Generation of Space Activities*, Studies in Space Policy 26, https://doi.org/10.1007/978-3-030-51559-1_5

5.1 Introduction

This chapter approaches the topic of on-orbit servicing (OOS) in relation to the African space sector by way of a critical question—is there room for the continent's space actors in this emerging field? In order to arrive at an answer to this question, the chapter begins by revisiting the instructive case of a relatively unknown effort to establish a Zambian space programme in the 1960s. This frames the rest of the chapter, and is followed by a brief review of the conceptualisation and evolution of OOS. Following this, the African space context is summarised, after which the importance of OOS for Africa is explored, as well as opportunities and obstacles for greater African participation in this field. Finally, the conclusion is presented, tying together all these aspects and presenting some key considerations.

In 1964, just as Zambia was achieving independence, Edward Mukuka Nkoloso, a grade-school science teacher and former freedom fighter, was leading an effort to launch Zambia's space programme at his unofficial National Academy of Science, Space Research, and Philosophy.[2] Nkoloso, who coined the term Afronauts, had the goal of beating the United States and the Soviet Union to the Moon, and then ultimately to reach Mars.[3] His space programme manifesto stated that "Our spacecraft, Cyclops I, will soar into deep abysmal space beyond the epicycles of the seventh heaven (…) Our posterity, the Black scientists, will continue to explore the celestial infinity until we control the whole of outer space".[4] However, this lofty ambition received no official support or funding despite a request for £7.000.000 from the United Nations Educational, Scientific, and Cultural Organisation (UNESCO) (other sources state variously $700 million and £700 million).[5] Nkoloso is also reported to have requested funds from Israel, the Soviet Union, the US, and the United Arab Republic, ranging in amount from "twenty million to two billion dollars".[6]

Despite his calls for support, in 1988 the Zambian Ministry of Power, Transport, and Communications publicly stated that "the matter under discussion was never seriously taken up by the Zambian Government and as such no official backing was rendered to Mr. Mukaka Nkoloso's efforts. The programme therefore died a natural death".[7] Given this lack of official support, Nkoloso's efforts generated only "a 10-rupee note sent by a space-minded Indian schoolboy".[8] Left with no alternative, he

---

[2]Lusaka Times, "Zambia's forgotten Space Program", 28 January 2011, https://www.lusakatimes.com/2011/01/28/space-program/.

[3]Namwali Serpell, "The Zambian 'Afronaut' Who Wanted to Join the Space Race", *The New Yorker*, 11 March 2017, https://www.newyorker.com/culture/culture-desk/the-zambian-afronaut-who-wanted-to-join-the-space-race.

[4]Ibid.

[5]Lusaka Times, "Zambia's forgotten Space Program".

[6]Serpell, "The Zambian 'Afronaut' Who Wanted to Join the Space Race".

[7]Lusaka Times, "Zambia's forgotten Space Program".

[8]Serpell, "The Zambian 'Afronaut' Who Wanted to Join the Space Race".

resorted to running his own makeshift training facility, and worked with a small group of volunteer trainees consisting of "twelve Zambian astronauts, including a (...) 16-year-old girl", whose training involved "spinning them around a tree in an oil drum and teaching them to walk on their hands, 'the only way humans can walk on the moon'".[9] The trainees would also "take turns to climb into an oil drum which rolled down a hill bouncing over rough ground".[10] He simulated zero gravity by "having them swing from the end of a long rope, cutting the rope when they reached the highest point so they went into freefall".[11] His plan was ultimately to send the "specially trained spacegirl, two cats (also specially trained) and a missionary" to Mars.[12] Despite Nkoloso's good intentions, he and his team were widely mocked and variously ridiculed as "crackpots",[13] and "a bit crazy",[14] while he was lambasted as "Zambia's village idiot".[15] Eventually, and despite his great dream for Zambia to be a participant in space, once Apollo 11 landed on the Moon "Nkoloso could not believe it, his hopes were shattered and his pursuance for his dream crashed. His trainees out of despair left the space academy".[16] In later years he pursued a law degree, graduating from the University of Zambia in 1983, and in one of his last interviews before his death in 1989, he continued to espouse his love for space: "I still have the vision of the future of man. I still feel man will freely move from one planet to another".[17]

In many ways, Nkoloso was a victim of his circumstances. During the struggle to free Zambia (formerly Northern Rhodesia) from British rule, he had been arrested, and subjected to torture at the hands of the Northern Rhodesian police. Some commentators argued, "after that, he lost it".[18] Furthermore, upon achieving independence, Zambia had "barely 1500 African-born high school graduates and less than 100 college graduates".[19] It was thus inconceivable to many that Zambia could participate in space or even have its own space programme, and today some view him as an inspiration since it "was just highly impossible to think like that in Zambia" at

---

[9]Alexis C. Madrigal, "Old, Weird Tech: The Zambian Space Cult of the 1960s", *The Atlantic*, 21 October 2010, https://www.theatlantic.com/technology/archive/2010/10/old-weird-tech-the-zambian-space-cult-of-the-1960s/64945/.

[10]China Central Television, "Faces of Africa 09/09/2013 Makuka Nkoloso: the Afronaut", 9 September 2013, https://web.archive.org/web/20170110112730/, http://english.cntv.cn/program/facesofafrica/20130909/100179.shtml.

[11]Lusaka Times, "Zambia's forgotten Space Program".

[12]Madrigal, "Old, Weird Tech: The Zambian Space Cult of the 1960s".

[13]Ibid.

[14]Lusaka Times, "Zambia's forgotten Space Program".

[15]Namwali Serpell, "The Afronaut Archives: Reports from a Future Zambia", *Public Books*, 28 March 2019, https://www.publicbooks.org/the-afronaut-archives-reports-from-a-future-zambia/.

[16]China Central Television, "Faces of Africa 09/09/2013 Makuka Nkoloso: the Afronaut".

[17]Serpell, "The Zambian 'Afronaut' Who Wanted to Join the Space Race".

[18]Ibid.

[19]Lusaka Times, "Zambia's forgotten Space Program".

the time.[20] As such, the purpose of this prologue is to argue that when approaching the topic of OOS from an African perspective, it is necessary for the continent to rekindle and fuel this visionary spirit of Edward Mukuka Nkoloso. His example shows clearly that Africans are not afraid to aspire to achieve greatness in space, and there is every reason to believe that this aspirational spirit also applies to this latest cutting-edge field. However, the continent must also learn valuable lessons from this example and properly support the visionary aspirations of Africans through the development of human capacity, technological and scientific skills, investments into research and development, and greater regulatory participation. Next, the conceptualisation and evolution of OOS will be presented.

5.2 On-Orbit Servicing: Conceptualisation and Evolution

OOS is regarded as a transformative and disruptive niche technology, and represents a major evolution in the establishment of an in-space economy. It entails the repair, refurbishment, refuelling, reuse, and recycling of satellites and other spacecraft, and "improves overall mission robustness and offers a unique capability to improve risk posture through post-launch operations".[21] Satellites could thus also be continuously upgraded as technologies evolve, greatly extending their lifespans. OOS also allows for the possibility of active debris removal, which is a critical consideration given the rapid growth in orbital debris. Of the approximately 9.600 satellites placed into orbit since 1957, about 5.500 are still in space, and about 2.300 are still active.[22] An estimated 34.000 objects larger than 10 cm are currently in orbit, with nearly a million objects ranging in size between 1 and 10 cm. Objects smaller in size than 1 cm number over a hundred million. In total, the mass of all space objects in Earth orbit is around 8.800 tonnes. It is thus with good reason that low-Earth orbit (LEO) is described as the "[w]orld's largest garbage dump".[23] No present or prospective space actor, whether developed, developing, commercial, governmental, military, or private, can thus afford to ignore the threat of space debris to the long-term sustainability of outer space activities.

---

[20] Kevin Aongola quoted in China Central Television, "Faces of Africa 09/09/2013 Makuka Nkoloso: the Afronaut".

[21] Mahashreveta Choudhary, "On-orbit satellite servicing: Process, Benefits and Challenges", *Geospatial World*, 27 July, 2018, https://www.geospatialworld.net/article/on-orbit-satellite-servicing-process-benefits-and-challenges-2/.

[22] European Space Agency, "Space debris by the numbers", February 2020, https://www.esa.int/Safety_Security/Space_Debris/Space_debris_by_the_numbers.

[23] National Aeronautics and Space Administration (NASA), "Space Debris", 1 July 2019, https://www.nasa.gov/centers/hq/library/find/bibliographies/space_debris.

Just like the space sector itself, OOS has evolved through several phases. Ushering in the first phase, the first true satellite servicing endeavour was the repair of the American Skylab station, which was launched in 1973. During the launch, the station's micrometeoroid shield, which was designed to protect the station from overheating, was damaged, one of the solar arrays was torn off, and the second solar array was jammed.[24] This severely limited the usefulness of the station, while also endangering any future crew. Consequently, during the first crewed mission, a spacewalk was undertaken to free the remaining solar array, and to deploy a makeshift parasol to keep the temperature in the station at acceptable levels. This first example of on-orbit servicing salvaged Skylab, and enabled a total of three crewed missions to conduct hundreds of experiments and gain valuable experience in living and working and space. A similar operation was the repair of Salyut 7 in 1985.[25] Contact with the station was lost and following a manual docking, major repairs had to be made to the station's electrical, water, and communications systems. Both of these examples are clear instances of OOS, and extended the lifespans of both stations. In the case of Salyut 7 in particular, the station was completely lost with no communications or power, but the bravery and skill of the Soviet crew salvaged it. This phase was thus characterised by governmental space agencies performing repairs on crewed space stations in orbit, and this legacy of human orbital repairs and servicing continues to the present, with the on-orbit construction of the International Space Station (ISS), and its regular maintenance, which also includes the use of spacewalks.

The second phase of OOS began in 1984. The Space Shuttle, first launched in 1981, opened up a range of new possibilities in terms of commercial servicing, salvaging, and recovery of satellites. One of the first satellites built from standardised components and designed for orbital repairs, Solar Maximum Mission (SMM), was launched in 1980 and later experienced failures in its attitude control system, which jeopardised its mission of observing the sun.[26] It also experienced problems with its coronagraph instrument electronics. In April, 1984, Space Shuttle Challenger was launched on a repair mission and after a rendezvous with the satellite, it was grabbed with the Remote Manipulator System (robotic arm), and after attaching it to the cargo bay's aft workbench, the attitude-control and electronics box were replaced.[27] SMM subsequently continued operations for five more years before re-entering the atmosphere in 1989. Famously, similar repairs have been made to the Hubble Space

[24]National Aeronautics and Space Administration (NASA), "Skylab 2: First Repair Spacewalk", 6 June 2018, https://www.nasa.gov/feature/skylab-2-first-repair-spacewalk.

[25]Nickolai Belakovski, "The little-known Soviet mission to rescue a dead space station", *Ars Technica*, 16 September 2014, https://arstechnica.com/science/2014/09/the-little-known-soviet-mission-to-rescue-a-dead-space-station/.

[26]Tracy McMahan and Valerie Neal, "Repairing Solar Max: The Solar Maximum Repair Mission", *National Aeronautics and Space Administration* (NASA), August 1984, https://ntrs.nasa.gov/archive/nasa/casi.ntrs.nasa.gov/19840020814.pdf.

[27]Randy Alfred, "April 11, 1984: Shuttle Makes House Call, Repairs Satellite", Wired, 11 April 2011, https://www.wired.com/2011/04/0411space-shuttle-astronauts-repair-solar-max-satellite/.

Telescope, which has been visited by five astronaut servicing missions.[28] Another milestone was reached when the Indonesian communications satellite Palapa B2, launched in 1984, failed to reach its intended geosynchronous orbit.[29] As a result of the failure, insurers had to pay $75 million to the Indonesian telecommunications agency, and subsequently took over ownership of the satellite.[30] They in turn contracted the National Aeronautics and Space Administration (NASA) to retrieve the satellite from orbit so it could be refurbished and relaunched. Palapa B2 was retrieved by the Space Shuttle Discovery later in 1984, and relaunched as Palapa B2-R in 1990, with the title transferred back to Indonesia.[31] During the same mission, another satellite which had also failed to reach its intended orbit, Westar 6, was also recovered and later relaunched as AsiaSat-1.[32] Unlike SMM, neither satellite was designed for orbital repair or recovery, so specific hardware had to be developed for the mission. It is clear that this second phase of OOS represented a landmark shift in the satellite and insurance industries, since, as was observed prior to the historic mission, the "first used satellite sale would help the insurers recover some of the money they paid after the failure".[33] This mission represented such a milestone that Lloyds of London awarded the crew its Silver Medal for Meritorious Service. The impact of satellite life extension cannot be overstated, and in another case, Intelsat 603, which was fitted with a boost motor by another Space Shuttle crew in 1992, which allowed it to reach its intended geosynchronous orbit, operated for a further 20 years and earned over $800 million. Unfortunately, with the end of the Space Shuttle programme in 2011, OOS capabilities were greatly hampered, with repairs to the ISS being the only current remaining exception.

However, another major evolutionary shift in OOS took place in 2019, when Mission Extension Vehicle 1 (MEV-1), built by Northrop Grumman, was launched atop a Soyuz rocket.[34] MEV-1 ushers in the next generation of OOS since it overcomes the need to have humans present in orbit to conduct the servicing of satellites. The first step of its first mission, to meet up with the 18-year-old Intelsat 901, was completed

---

[28] National Aeronautics and Space Administration (NASA), "About—Hubble Servicing Missions", 7 February 2020, https://www.nasa.gov/mission_pages/hubble/servicing/index.html.

[29] Sattel Technologies, "The Historic Journey of Palapa B2", no date, https://www.sattel.com/life_of_palapa_b2.htm.

[30] John Noble Wilford, "Nasa Plans to Use Shuttle to Retrieve a Satellite", *The New York Times*, 17 August 1984, https://www.nytimes.com/1984/08/17/us/nasa-plans-to-use-shuttle-to-retrieve-a-satellite.html.

[31] Sattel Technologies, "The Historic Journey of Palapa B2".

[32] Richard Parker, "On-orbit satellite servicing, insurance and lessons of Palapa B2 and Westar 6", *Room*, 2015, https://room.eu.com/article/Onorbit_satellite_servicing_insurance_and_lessons_of_Palapa_B2_and_Westar_6.

[33] Wilford, "Nasa Plans to Use Shuttle to Retrieve a Satellite".

[34] Darrell Etherington, "The first spacecraft that can service satellites to extend their life launched today", Tech Crunch, 9 October 2019, https://techcrunch.com/2019/10/09/the-first-spacecraft-that-can-service-satellites-to-extend-their-life-launched-today/.

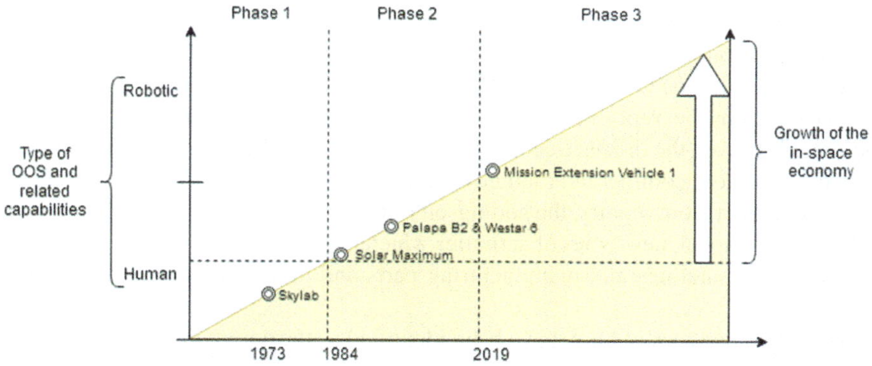

Fig. 5.1 Evolution of on-orbit servicing

in February 2020.[35] MEV-1 has been contracted to provide support for the satellite, which is running low on propellant, for a period of five years, after which it will push the satellite into a decommissioning orbit to allow another satellite to take over its slot in geosynchronous orbit. MEV-1 is designed to provide more than 15 years of life-extension services, and is capable of docking and undocking with multiple vehicles. Northrup Grumman has stated that its vision is to have "a fleet of satellite-servicing vehicles that not only extend the life of satellites, but provide other services such as inclination changes and spacecraft inspections, as well as use advanced robotics technology to perform additional functions such as in-orbit repair and assembly".[36] While MEV-1 is essentially a space tug, "even that can as much as double the life of some geosynchronous satellites—which means a lot more potential revenue for not too much more cost".[37] It clearly represents a "new era for commercial satellite operation, leading to further decreased operating costs, and therefore more access for startup and smaller companies to take part".[38] A second MEV is planned for later in 2020, while NASA is set to launch its Restore-L robotic refuelling demonstration soon (now known as OSAM-1 for On-Orbit Servicing, Assembly and Manufacturing mission 1). The US Defence Advanced Research Projects Agency (DARPA) is also working on the robotic servicing of geosynchronous satellites. Orbit Fab is another company working in the field, and is focusing on its "gas stations in space" approach to simple refuelling of spacecraft without the need to dock. Other potential future opportunities for this new phase of OOS include upgrading the capabilities of satellites to maximise return on investment.

Figure 5.1 depicts the evolution of OOS up to the present, including major milestones ushering in each new phase. The Solar Maximum servicing and Palapa B2 and

[35]Elizabeth Howell, "Two private satellites just docked in space in historic first for orbital servicing", Space.com, February 2020, https://www.space.com/private-satellites-docking-success-northrop-grumman-mev-1.html.

[36]Ibid.

[37]Etherington, "The first spacecraft that can service satellites to extend their life launched today".

[38]Ibid.

Westar 6 recovery, all in 1984, ushered in the new era of commercial OOS activities, as opposed to purely governmental OOS. The recent addition of MEV-1 ushers in the new phase of robotic OOS. However, as Fig. 5.1 shows, this does not replace human OOS operations, but represents an expansion of capabilities and builds on what came before. Moreover, the in-space economy is set to see a rapid expansion, with the addition of even more commercial (and governmental and military) providers of robotic OOS, and apart from repairs, the adding-on of new capabilities to existing satellites, and debris removal, new types of activities which include remotely reusing and recycling defunct satellites and manufacturing parts and new craft can soon become a reality.

Five specific types of OOS have been identified as the commercially "most interesting" missions.[39] These are on-orbit refuelling, attitude and orbital control systems (AOCS) life extension, deorbit, array operations (where the servicing craft attaches to a satellite to provide it with power), and mechanical intervention (assisting with equipment not correctly deployed). Of the milestone examples above, Skylab, Salyut 7, and the ISS are all examples of mechanical interventions, SMM was an instance of AOCS life extension and mechanical intervention (since parts were replaced), Palapa B2 and Wester 6 were examples of deorbit, and MEV-1 is providing AOCS life extension by providing station-keeping and attitude control manoeuvres. However, these types of OOS exclude other potential scenarios such as on-orbit refurbishment and transfer of ownership, and salvaging parts from defunct satellites to use as replacement parts or to build new satellites, not to mention the possibility of recycling existing parts to manufacture new ones on-orbit. All of these present the potential to transform the space sector, and to pose disruptive challenges to the industry.

In particular, they pose significant legal challenges. While these will not be discussed in depth here, the key point is that this new generation of space activities was unforeseen by the drafters of the five core space treaties, since the "existing international body of law was originally created with state civilian or military actors in mind and therefore lacks the specificity and legal certainty that is necessary for mature commercial activities".[40] In particular, various legal questions will have to be answered in relation to the Outer Space Treaty (OST) and the Liability Convention (LIAB). As a first point, the OST (Art. VIII) stipulates that "A State Party to the Treaty on whose registry an object launched into outer space is carried shall retain jurisdiction and control over such object, and over any personnel thereof, while in outer space or on a celestial body".[41] However, as the data on space debris and objects

[39] Martin J. Losekamm, Jacob Hacker, Nikita Sardesai, Anja Nakarada Pecujlic, and Adam Vigneron, "Legal and Political Implications of Future On-Orbit Servicing Missions", *66th International Astronautical Congress*, Jerusalem, Israel, 2015, 3, https://www.researchgate.net/publication/282979175_Legal_and_Political_Implications_of_Future_On-Orbit_Servicing_Missions.

[40] Ibid., 5.

[41] United Nations Office for Outer Space Affairs, "International Space Law: United Nations Instruments", 2017, 6, https://www.unoosa.org/res/oosadoc/data/documents/2017/stspace/stspace61rev_2_0_html/V1605998-ENGLISH.pdf.

in space showed earlier, with so many defunct satellites remaining in orbit, and with so many pieces of debris, identifying the launching state and owner of all objects is essentially impossible. However, if debris is to be cleared through OOS, permission will have to be obtained from the relevant owner of each object before it can be deorbited. This presents a major complication since "The ownership of an asset in space remains with the state it is registered with practically forever, with no such thing as a space salvage law—analogous to the right to salvage in maritime law—existing. Consent has to be obtained on a case-by-case basis, by specific agreements between state actors".[42] The second major concern is related to liability for damages, since the Liability Convention (Art. II)[43] stipulates that "A launching State shall be absolutely liable to pay compensation for damage caused by its space object on the surface of the Earth or to aircraft in flight"—as such, even if permission could be granted for debris to be deorbited through OOS, the launching State remains liable for resulting damages. Article III states that "In the event of damage being caused elsewhere than on the surface of the Earth to a space object of one launching State or to persons or property on board such a space object by a space object of another launching State, the latter shall be liable only if the damage is due to its fault or the fault of persons for whom it is responsible".[44] The issue is further complicated by the stipulation (LIAB, Art. IV-1) that "In the event of damage being caused elsewhere than on the surface of the Earth to a space object of one launching State or to persons or property on board such a space object by a space object of another launching State, and of damage thereby being caused to a third State or to its natural or juridical persons, the first two States shall be jointly and severally liable to the third State".[45] The confusion resulting from these stipulations is clear:

> These formulations leave room for interpretation. It is not clear which state is liable in the event that an accident occurs during a servicing or active debris removal mission, possibly involving a servicer spacecraft of yet another launching state. Who would be liable for damage caused on Earth by a space object being deorbited during an active removal mission? One could argue that according to Article III of the Liability Convention the owner state of the deorbited spacecraft may not be liable, but Article III specifically refers to damage caused "elsewhere than on the surface of Earth". Yet Article II states that the launching state is always liable for damage caused on the surface of Earth. Does that apply even when the spacecraft was deorbited by an actor other than the launching state?[46]

Finally, there is also no binding regulation in terms of either the mitigation of debris or its active removal. The point here is that commercial operators of OOS are exposed to great financial risks because of the lack of clarity contained in the legal code. It is also clear that much work must be done to clarify these issues before OOS

---

[42] Losekamm, Hacker, Sardesai, Nakarada Pecujlic, and Vigneron, "Legal and Political Implications of Future On-Orbit Servicing Missions", 5.

[43] United Nations Office for Outer Space Affairs, "International Space Law: United Nations Instruments", 15.

[44] Ibid.

[45] Ibid.

[46] Losekamm, Hacker, Sardesai, Nakarada Pecujlic, and Vigneron, "Legal and Political Implications of Future On-Orbit Servicing Missions", 5.

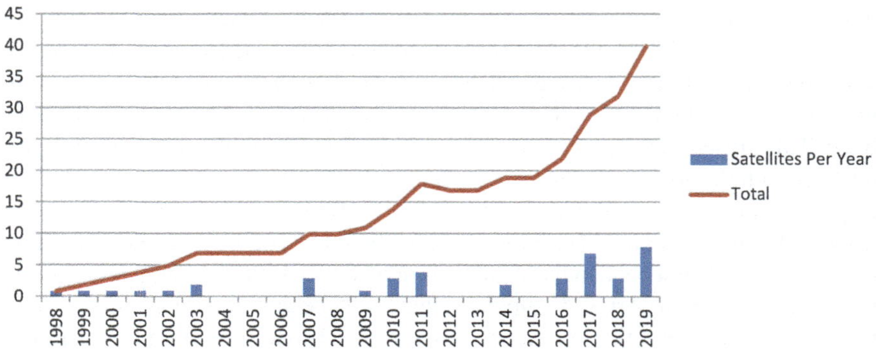

Fig. 5.2 African satellites

can fulfil its revolutionary potential. In the next section, the African space sector is reviewed with a focus on the continent's participation in space activities.

5.3 The African Context

Despite Edward Mukuka Nkoloso's efforts, the African space sector was only truly established after the end of the Cold War.[47] Since the start of the twenty-first century, the sector has rapidly grown, and as Fig. 5.2 shows, a total of 40 African satellites have been launched by 2020. This number has been rapidly increasing in recent years as more emerging space actors across the continent take advantage of the opportunities offered by small and relatively inexpensive standardised satellites and components, as well as the growth in collaborative space projects across the continent and globe. Three satellites have also been launched by African partners collaborating in multilateral projects, namely RascomStar (the Regional African Satellite Communication Organisation, a telecommunications operator registered in Mauritius, which has launched two satellites) and New Dawn (a communications satellite which was funded primarily by South African investors including Nedbank, the Industrial Development Corporation of South Africa, the African Development Bank and the investment management firm Convergence Partners[48]). However, in the latter case, these investors sold their stake in New Dawn to Intelsat in 2012, after which it became

[47]For a full discussion, see: Annette Froehlich and André Siebrits, *Space Supporting Africa Volume 1: A Primary Needs Approach and Africa's Emerging Space Middle Powers* (Cham, Switzerland: Springer, 2019).

[48]Brand South Africa, "Africa's New Dawn satellite in orbit", 4 May 2011, https://www.brandsouthafrica.com/investments-immigration/science-technology/intelsat-newdawn.

the sole owner and operator of the satellite.[49] As a consequence of this transfer of ownership, the total number of African satellites drops by one in 2012 in Fig. 5.2.

The origins of these satellites are depicted in Fig. 5.3, illustrating the diverse range of participation in the sector. While some of these satellites have been built by traditional space agencies (with or without partners), some such as GhanaSat-1 was built and tested by students from the All Nations University (Koforidua) in partnership with the Kyushu Institute of Technology and its Joint Global Multi-Nation Birds Satellite project (or BIRDS for short).[50] Costing about half a million dollars, no Ghanaian public money was used in this project. This proves that Africans are finding innovative ways to participate in space activities, and that the range of those activities is diversifying. For an in-depth review of African satellites, see Froehlich and Siebrits.[51]

One of the challenges related to this rapid increase in African space activities, is the uneven and incomplete implementation of the five outer space treaties by African countries, especially those undertaking space activities. For example, as of 1 January 2019, Egypt has signed but not ratified the Liability Convention and has not signed the Registration Convention, Ethiopia has signed but not ratified the Outer Space Treaty (and has not signed any other), Kenya has not signed or ratified the Registration Convention, Rwanda and Ghana have only signed but not ratified the Outer Space Treaty, Rescue Agreement, and Liability Convention, and Angola and Sudan have signed or ratified none.[52] Morocco is the only African country to have ratified all five treaties. There is thus a clear deficit in regulatory participation and action by active spacefaring countries, which is a matter of concern. Moreover, as of 2019, only 20 of the 54 African states have joined the United Nations Committee on the Peaceful Uses of Outer Space (Fig. 5.4). The concern here is that, as a result, less than half of African states are participating in critical debates around issues such as OOS, space debris mitigation, and others. For example, the report on the 2020 session of the Scientific and Technical Subcommittee (STSC) of UNCOPUOS stated that "[t]he Subcommittee noted a number of measures that had been or were being undertaken to implement the Guidelines for the Long-term Sustainability of Outer Space Activities. Those measures include … the development of standards for on-orbit servicing and

[49] Jeffrey Hill, "African New Dawn Investors Sell Joint Venture Share Back to Intelsat", *Via Satellite*, 9 November 2012, https://www.satellitetoday.com/telecom/2012/11/09/african-new-dawn-investors-sell-joint-venture-share-back-to-intelsat/.

[50] Louis de Gouyon Matignon, "GhanaSat-1, the first Ghanaian satellite", *Space Legal Issues*, 26 March 2019, https://www.spacelegalissues.com/space-law-ghanasat-1-the-first-ghanaian-satellite/.

[51] Annette Froehlich and André Siebrits, *Space Supporting Africa Volume 1: A Primary Needs Approach and Africa's Emerging Space Middle Powers* (Cham, Switzerland: Springer, 2019).

[52] United Nations Committee on the Peaceful Uses of Outer Space, "Status of International Agreements relating to activities in outer space as at 1 January 2019", 1 April 2019, https://www.unoosa.org/documents/pdf/spacelaw/treatystatus/AC105_C2_2019_CRP03E.pdf.

Algeria	AlSat-1 (2002) AlSat-2A (2010) AlSat-1B (2016) AlSat-2B (2016) AlSat-1N (2016) AlComSat-1 (2017)
Angola	AngoSat-1 (2017)*
Egypt	NileSat-101 (1998) NileSat-102 (2000) EgyptSat-1 (2007)* NileSat-201 (2010) EgyptSat-2 (2014)* EgyptSat-A (2019) NARSSCube-1 (2019) NARSSCube-2 (2019) TIBA-1 (2019)
Ethiopia	ETRSS-1 (2019)
Ghana	GhanaSat-1 (2017)
Kenya	1KUNS-PF (2018)
Morocco	Maroc-TUBSAT (2001) Mohammed VI-A (2017) Mohammed VI-B (2018)

Nigeria	NigeriaSat-1 (2003) NigComSat-1 (2007)* NigeriaSat-2 (2011) NigeriaSat-X (2011) NigComSat-1R (2011) NigeriaEduSat-1 (2017)
Rwanda	RwaSat-1 (2019)
South Africa	SUNSAT (1999) ZACube-1 (2003) SumbandilaSat (2009) KONDOR-E (2014) nSight1 (2017) ZA-AeroSat (2017) ZACube-2 (2018) XinaBox ThinSat (2019)
Sudan	SRSS-1 (2019)
Other	RascomStar-QAF 1 (2007)* RascomStar-QAF 1R (2010) New Dawn (2011) [later Intelsat 28]

Fig. 5.3 African satellites by country/entity (* indicates failed or impaired satellites)

Fig. 5.4 African Member States of the United Nations Committee on the Peaceful Uses of Outer Space (as of 2019). United Nations Office for Outer Space Affairs, "Committee on the Peaceful Uses of Outer Space: Membership Evolution", 2020, https://www.unoosa.org/oosa/en/ourwork/copuos/members/evolution.html

rendezvous and proximity operations"[53] and that "[t]he Subcommittee noted the evolving technologies related to the in-orbit robotic servicing of satellites and the extension of satellite lifespans"[54]—yet the only African states present at the 2020 STSC were Algeria, Egypt, Kenya, Libya, Morocco, Nigeria, South Africa, and

[53] United Nations General Assembly, "Draft report: X. Long-term sustainability of outer space activities", *Committee on the Peaceful Uses of Outer Space Scientific and Technical Subcommittee*, Fifty-seventh session, 12 February 2020, 3, https://www.unoosa.org/res/oosadoc/data/documents/2020/aac_105c_1l/aac_105c_1l_385add_5_0_html/AC105_C1_L285Add05E.pdf.

[54] United Nations General Assembly, "Draft report: I. Introduction", *Committee on the Peaceful Uses of Outer Space Scientific and Technical Subcommittee*, Fifty-seventh session, 7 February 2020, 6, https://www.unoosa.org/res/oosadoc/data/documents/2020/aac_105c_1l/aac_105c_1l_385add_1_0_html/AC105_C1_L285Add01E.pdf.

Tunisia.[55] Similarly, at the 2019 Legal Subcommittee session, only Algeria, Burkina Faso, Egypt, Kenya, Libya, Morocco, Nigeria, and South Africa were in attendance.[56]

This is a troubling image since it effectively means that a few states and their representatives are making the rules for everyone, including the rules for OOS. In this regard, it is true that "Industrialized and powerful established space nations have the upper hand when it comes to the setting of international standards, rules and norms applicable to space activities" resulting in "asymmetric development of international space industry regulations and standards … [stemming] from the bias that is exhibited in favour of powerful established space nations that are active in the standard setting process and who influence standardization".[57] Two critical dangers can result from this bias: "With limited participation from emerging space nations in the formation of international space regulations and standards, the resulting laws and rules may be regarded by such States as unrepresentative or invalid" and "the less active emerging space nations are in developing their own space standards, the more likely it may be that these States will be at risk of being sidelined".[58] As such, it is critical that more African states have a seat at the table when it comes to the development of regulations and standards, most critically at the highest levels such as UNCOPUOS, and that those states that already have a seat at the table occupy that seat and participate actively. The critical observation in this regard is that "it is not enough to invest in technology. One must also have the capacity to understand and shape regulatory agendas around technologies".[59] Must greater effort must also be made in this regard in the fostering and development of African legal and policy-making skills in the space sector. One powerful suggestion is for the establishment of an equivalent institution in Africa of the European Space Policy Institute.[60] This is especially critical since "Space Law is almost non-existent in the academic curricula of universities across the continent, except for a few universities that offer space studies as a unit of postgraduate programme".[61] As a consequence, there is a dearth

---

[55]United Nations General Assembly, "List of Participants", *Committee on the Peaceful Uses of Outer Space Scientific and Technical Subcommittee*, Fifty-seventh session, 13 February 2020, https://www.unoosa.org/res/oosadoc/data/documents/2020/aac_105c_1inf/aac_105c_12020inf49_0_html/AC105_C1_INF_2020_049EFS.pdf.

[56]United Nations General Assembly, "List of Participants", *Committee on the Peaceful Uses of Outer Space Legal Subcommittee*, Fifty-eighth session, 11 April 2019, https://www.unoosa.org/res/oosadoc/data/documents/2019/aac_105c_2inf/aac_105c_22019inf51_0_html/AC105_C2_2019_INF51EFS.pdf.

[57]Joel A. Dennerley, "Emerging Space Nations and the Development of International Regulatory Regimes", *Space Policy* 35, (2016): 28.

[58]Ibid., 28–29.

[59]Ibid., 30.

[60]Etim Offiong and Valanathan Munsami, "Towards a space policy institute for Africa," *Space Policy* 46, (2018).

[61]Joseph Ibeh, "Towards Developing Africa's Think Tank In Space Law and Policy", *Space in Africa*, 18 July 2019, https://africanews.space/towards-developing-africas-think-tank-in-space-law-and-policy/.

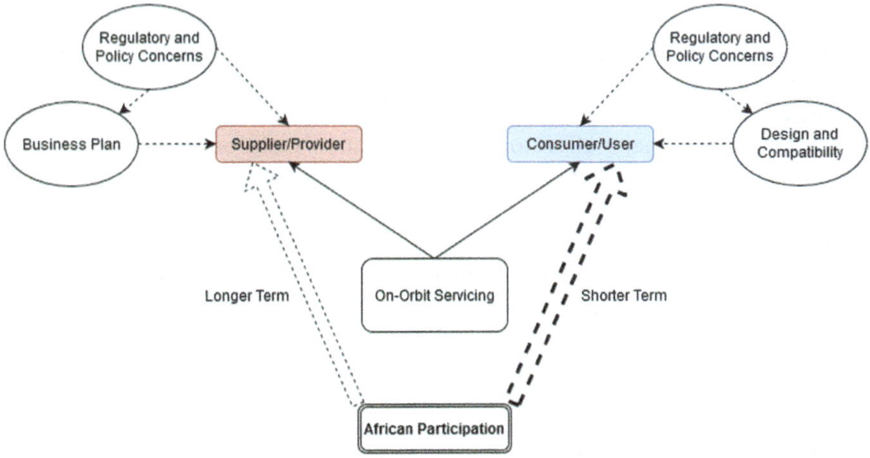

Fig. 5.5 African on-orbit servicing equation

of expertise when it comes to drafting critical documents and analysing space-related policies. As such, if there is to be room for Africa in the field of OOS (especially given the related legal questions and concerns that must be resolved) and if the continent is to protect its own interests, "Africa must move beyond mere scepticism at proposals that seek to make change, but be proactive about contributing ideas about what that change should look like".[62] In some respects, it unfortunately remains true, as Edward Mukuka Nkoloso remarked of Zambia in 1964, that "Most Westerners don't even know whereabouts in Africa we are"[63]—the same applies to space-related regulatory and policy spheres for far too many African states today.

5.4 African On-Orbit Servicing Needs, Opportunities, and Obstacles

Like any commercial sector, OOS consists of two closely-related dimensions—demand and supply. As such, this section will begin by considering whether there is room for Africa on the demand side of the OOS equation (Fig. 5.5). Of the 40 African satellites shown in Fig. 5.3, five have had critical failures or have otherwise been impaired. Angola's first (Russian-built—RKK Energia) satellite, AngoSat-1,

[62]Timiebi Aganaba-Jeanty, "Why Africa must move beyond scepticism to influence international law", *Business Day*, 26 August 2014, https://businessday.ng/analysis/article/why-africa-must-move-beyond-scepticism-to-influence-international-law/.

[63]Serpell, "The Afronaut Archives: Reports from a Future Zambia".

which was a communications satellite built at a cost of about $330 million, experienced communications problems and was declared lost only four months after launch.[64] While a new, more powerful replacement satellite is being built at no cost to Angola, the loss of the country's first national communications spacecraft has been a blow to the "major effort to revamp the nation's telecommunications infrastructure and switch local TV channels from analog to digital format ... [and] to improve and expand communications across all 117 municipalities and to link the nation to the rest of the African continent".[65] Nigeria's first (Chinese-built—China Great Wall Industry Corporation) communications satellite, NigComSat-1, was lost less than two years after launch, due to failures of its two solar arrays.[66] The cost of this satellite was approximately $300 million, and due to insurance a replacement was built and launched in 2011.[67] The loss of NigComSat-1 was a major blow to Nigeria's efforts to reduce its leasing of capacity aboard foreign satellites, which cost the country $1 billion per year at the time. The four-year delay before the replacement satellite could enter service thus had very real economic consequences. RascomStar-QAF 1, the first communications satellite of Rascom, built by Alcatel Alenia Space, suffered a propulsion malfunction in the form of a helium leak, which reduced its effective lifespan to only two years.[68] Libya contributed $300 million to the satellite.[69] A replacement was launched three years later. EgyptSat-1 (Ukrainian-built Earth observation satellite) stopped communicating three years after launch,[70] and while EgyptSat-2 (Russian-built Earth observation satellite with an 11-year planned lifespan) was launched in April, 2014, full control was handed over to the Egyptian National Authority for Remote Sensing and Space Sciences (NARSS) in January, 2015, and was lost in June, 2015.[71] Since the satellite had been handed over to Egypt, "The Russian side cannot be blamed for the malfunction, as EgyptSat-2 started experiencing technical problems after it was put under full control of Egyptian specialists", and as a result Egypt was "considering purchasing Russian satellite images of

---

[64] Al Jazeera News, "Angola's first satellite defunct four months after launch", 24 April 2018, https://www.aljazeera.com/news/2018/04/angola-satellite-defunct-months-launch-180424072213475.html.

[65] Anatoly Zak, "The Angosat-1 communications satellite", *Russian Space Web*, 17 February 2020, https://www.russianspaceweb.com/angosat.html.

[66] Gunter Dirk Krebs, "NigComSat 1, 1R", *Gunter's Space Page*, 21 July 2019, https://space.skyrocket.de/doc_sdat/nigcomsat-1.htm.

[67] Peter B. de Selding, "Nigcomsat-1R Launched Successfully by Long March", *Space News*, 21 December 2011, https://spacenews.com/nigcomsat-1r-launched-successfully-long-march/.

[68] Gunter Dirk Krebs, "Rascom-QAF 1, 1R", *Gunter's Space Page*, 11 December 2017, https://space.skyrocket.de/doc_sdat/rascom-1.htm.

[69] Tortilla Con Sal, Satellites, "False Beliefs and Sovereign Integration", *La nueva Televisión del Sur*, 3 September 2015, https://www.telesurenglish.net/bloggers/Satellites-False-Beliefs-and-Sovereign-Integration-20150903-0001.html?fb_comment_id=1010410769001640_1017828864926497.

[70] European Space Agency, "EgyptSat-1", *eoPortal Directory*, 2020, https://directory.eoportal.org/web/eoportal/satellite-missions/e/egyptsat-1.

[71] European Space Agency, "EgyptSat-2", *eoPortal Directory*, 2020, https://directory.eoportal.org/web/eoportal/satellite-missions/e/egyptsat-2.

Earth".[72] In all these cases, Africa was left without vital services, and if OOS had been available it could have saved both time and money for all involved. It is also not difficult to see why insurance costs can be so high in the space sector. While all satellites ultimately fail due to the harsh space environment, these were some of the clearest examples illustrating Africa's short-term OOS needs.

In the African OOS Equation, the sector is broken down into its supplier/provider and consumer/user dimensions. Both of these are equally impacted by regulatory and policy concerns, for example around liability issues as discussed earlier, and as such, potential users of OOS also have a direct interest in participating and shaping the discourse and further development of international space law and related policies. On the supply side, the technology is available to successfully provide on-orbit services, such as those mentioned in Sect. 5.2. The major obstacle for the provision of OOS (apart from the legal concerns also mentioned earlier) has been economic feasibility—"What has been lacking is really the business plan from the commercial operators to justify moving forward … What hasn't been lacking is the technology to be able to do it".[73] From the demand side, a major concern is designing satellites to be compatible with OOS:

> Preparing a satellite to be serviced means not only including in its design a 'grapple fixture' and an interface to allow for fuel transfer, but the satellite must also be in a cooperative state during rendezvous and capture by the service provider … Furthermore, the serviced spacecraft could also require dedicated markers or patterns in specific locations, or it may need new attitude and pointing abilities to make rendezvous and capture possible. Thus, there is the need to define guidelines for future telecom satellites which describe all of the changes required to enable servicing and refuelling.[74]

While space-related capabilities are unevenly distributed across Africa (with South Africa, Algeria, Egypt, and Nigeria being arguably the most advanced), the African space sector as a whole is still in a relatively nascent stage. As such, over the shorter term, African involvement in OOS will likely be predominantly on the demand side. This does not however mean that the continent's space actors can afford to ignore the need to participate in establishing guidelines, policies, and laws pertaining to the sector. There is also a critical need to incorporate compatible designs into future satellites to allow for OOS as the need arises. This has to be done in advance however, because once the satellite is in orbit it is too late to make modifications. This means there is a critical need for OOS to be at the top of the agenda across African space-related fora, including in the African Union. Since "Satellites are incredibly expensive to manufacture, and launch costs can be astronomical … [and] much of the risk lies in launching the craft to space", it is in the interest of all African space sector participants to take OOS seriously and protect the continent's investments.

[72] Ibid.

[73] Swapna Krishna, "The Time For On-Orbit Satellite Servicing is Here", *Via Satellite*, August 2018, https://interactive.satellitetoday.com/via/august-2018/the-time-for-on-orbit-satellite-servicing-is-here/.

[74] European Space Agency, "On-Orbit Servicing Prepared Platforms", 23 September 2019, https://blogs.esa.int/cleanspace/2019/09/23/on-orbit-servicing-prepared-platforms/.

Satellites serve critical needs across multiple sectors on the continent,[75] and failures or other impairments have serious and direct economic, social, and environmental consequences.

Over the longer-term, especially if the African Space Agency (AfSA) is a success as intended, it is possible that African space actors could also enter the OOS equation on the supply side. AfSA was founded in 2018, and according to its statute, the agency shall "[h]arness the potential benefits of space science, technology, innovation and applications in addressing Africa's socio-economic opportunities and challenges", while "[d]evelop[ing] a sustainable and vibrant indigenous space market and industry that promotes and responds to the needs of the African continent" and "[s]trengthen[ing] space missions on the continent in order to ensure optimal access to space-derived data, information, services and products".[76] Egypt was selected as the host of the Agency in 2019, and is building the headquarters in New Cairo.[77] As seen with the expensive satellite failures mentioned above, OOS would certainly serve the 'needs of the African continent' while also supporting continued access to space-derived services and data. While OOS is not mentioned in the AfSA statute, it is strongly recommended that the agency considers becoming an OOS provider in the future. Given the expected growth of the in-space economy in coming years, building African capacity in this regard serves the goal of growing a 'vibrant indigenous space market and industry'. However, while there are significant opportunities in this regard, there are also serious challenges. Table 5.1 presents a brief initial consideration of African involvement in OOS on both demand and supply sides via a strengths, weaknesses, opportunities, and threats (SWOT) analysis. It is recommended that this analysis is continued in future research.

5.5 Lessons from the Past: Nkoloso's Ghost and Africa's Future in On-Orbit Servicing

In order to answer the question of whether there is space for Africa in the innovative emerging sector of OOS, this chapter began by considering the case of Edward Mukuka Nkoloso and his dream to establish Zambia's space agency. This was done

---

[75] For in-depth discussions of the role and importance of satellites for Africa, see: Annette Froehlich (ed.), *Embedding Space in African Society: The United Nations Sustainable Development Goals 2030 Supported by Space Applications* (Cham, Switzerland: Springer, 2019); Annette Froehlich (ed.), *Integrated Space for African Society: Legal and Policy Implementation of Space in African Countries* (Cham, Switzerland: Springer, 2019); Annette Froehlich (ed.), *Space Fostering African Societies: Developing the African Continent through Space, Part 1* (Cham, Switzerland: Springer, 2019).

[76] African Union, "Statute of the African Space Agency", 29 January 2018, 3, https://au.int/sites/default/files/treaties/36198-treaty-statute_african_space_agency_e.pdf.

[77] Egypt Today, "Egypt allocates $10 mn to establish African Space Agency", 23 February 2019, https://www.egypttoday.com/Article/1/65133/Egypt-allocates-10-mn-to-establish-African-Space-Agency.

Table 5.1 African OOS SWOT Analysis

Africa as consumer/user of OOS		Africa as provider/supplier of OOS	
Strengths The African space sector is growing rapidly, and more actors across the continent are building and launching satellites. Thus, Africa is strongly positioned to be a consumer of OOS	*Weaknesses* There are significant gaps in regulatory participation and policymaking. Issues of liability, space debris, and malicious interference with African satellites are all concerns that must be addressed via the appropriate fora	*Strengths* Via the African Space Agency, African Space Policy and Strategy, and other instruments, Africa is taking steps to foster a strong space sector. Since AfSA will represent and draw on talents across the entire continent, it could have significant capabilities	*Weaknesses* Lack of funding, skills shortages, remaining scepticism and apathy towards space activities, uneven space sector development and participation across Africa
Opportunities Several potential suppliers are looking into supplying OOS services (ex. Northrup Grumman), thus if proactive measures are taken to design satellites to be compatible with OOS, the costs of failures/insurance could go down and delays/dependence on foreign space actors for data could be avoided	*Threats* If proactive steps are not taken to ensure satellites are compatible with OOS, it will be of no use to the continent. The shortage of skills and human capacity is a potential threat to the entire African space sector	*Opportunities* While some commercial, military, and governmental actors globally are taking steps to participate in OOS, the sector is still new and there is room for new entrants in the provision of OOS services. The in-space economy could also become a significant commercial sector	*Threats* Lack of African representation and participation in regulatory affairs. This could open up the possibility of others making rules for the OOS sector, which African actors must follow later. Not paying sufficient attention to OOS and the skills development needed for taking Africa's place in this emerging field

for a good reason. Just like the space sector as a whole in the early to middle 1960s, the OOS sector is currently in its nascent stage and the potential exists to transform the satellite industry—an industry that Africa is increasingly taking part in today. While the analysis of the historical evolution of OOS showed that the concept itself is not new, it was previously the exclusive domain of developed governmental space actors with the most advanced space capabilities. With the launch of MEV-1 by a commercial space actor, not only is a new evolutionary phase ushered in through the use of robotic servicing craft, but the domain is opening up to include more diverse actors (this is part of the general evolution of the space sector as a whole in the form of NewSpace and Space 4.0). It is in this context that Africa is now becoming a much more active player in space, and many more satellites are set to be launched by the continent in the coming years. Space is becoming more democratised and accessible as technologies are becoming miniaturised and standardised, with commercial off-the-shelf components offering more capabilities than ever before.

So why is it necessary to look back to the example of Nkoloso in order to look forward to the future?

First, Nkoloso demonstrated that Africans can pursue ambitious space dreams and given the context of Zambia at the time, it was astonishing that someone could 'think big' and be daring enough to challenge the superpowers in their space race. Africa needs this daring spirit now more than ever. There is no room for defeatist attitudes in the space sector. As arguments proffered by various scholars showed, if Africans are not taking their place at the table, they will simply be sidelined and the rules will be made by others. It is worrying that space actors are not taking up the responsibility of acceding to critical space treaties, joining in discussions on the future of the space sector at the highest levels, or simply not showing up during meetings. Second, Nkoloso and his space cadets were widely ridiculed for their apparently amateurish space efforts, but given their total lack of formal support it is difficult to imagine any other outcome. The fact that Nkoloso later earned a law degree shows that there was clear potential, and that he was not the 'village idiot' of Zambia he was made out to be. This example is a painful lesson for all of Africa. Even today, while there are some exceptions, "the data shows that countries in sub-Saharan Africa—with the exception of South Africa, Kenya and Senegal—spend only 0.4% of their gross domestic product (GDP) on research", while the "brain drain continues to hurt African countries and African science, partly, because many scientists are not well paid".[78] In 2017, it was observed that "the total R&D expenditure of the region [African continent] falls below 1% of the GDP".[79] As argued elsewhere,[80] there is a lingering scepticism towards the space sector in Africa, and space activities are seen as being in competition to other, more pressing priorities. However, there is often too little appreciation of the role of space technologies in supporting and promoting African development in precisely these other priority areas, since space plays a critical role in increasingly numerous sectors of our daily lives. While the establishment of (some) African national space agencies and policies, as well as the continental space agency and policy/strategy are encouraging signs that the continent is taking space more seriously, the risk still exists that far too many initiatives and activities are 'never seriously taken up', receive little or 'no official backing', and are at risk of 'a natural death'—thus echoing the experiences of Nkoloso and his space academy. Poor performance and investment in critical STEM (science, technology, engineering, and mathematics) subjects is another major obstacle.[81] However, the

[78]Esther Ngumbi, "African governments must invest in science for future growth", *Mail & Guardian*, 23 November 2019, https://mg.co.za/article/2019-11-23-00-african-governments-must-invest-in-science-for-future-growth/.

[79]Samuel Oyewole, "Space Research and Development in Africa", *Astropolitics* 15, no. 2 (2017): 196.

[80]See: André Siebrits, "Reflective Practice in the African Space Sector: The Importance of Cadre Formation", in *Space Fostering African Societies: Developing the Continent through Space, Part 2* (Cham, Switzerland: Springer) *forthcoming*.

[81]For a more in-depth discussion, see in particular the section on education in: Annette Froehlich and André Siebrits, *Space Supporting Africa Volume 1: A Primary Needs Approach and Africa's Emerging Space Middle Powers* (Cham, Switzerland: Springer, 2019); For a review of the tertiary

STEM subjects are not the only critical ones, and it is vital that natural scientists and social scientists should be engaged in interdisciplinary dialogue: "Social scientists recognise that the human element of science can never be fully absent, and so they attempt to include the human, subjective perspective in their quest for answers. Understanding each other's methods helps to put our own in perspective".[82]

Since OOS offers to transform the space industry as a whole, it is in the interest of all Africans that we do not ask 'what could have been' in the decades to come, as in the case of Nkoloso and his space academy. Legal questions of grave concern for the continent are under debate, and Africans of all nationalities must take greater part in these. Issues such as space debris mitigation, the sustainability of outer space activities, liability, and ownership have direct consequences to the continent's ambitions to develop a 'sustainable and vibrant indigenous space market and industry'. A critical step in the engagement of the continent's space actors with OOS is to be good consumers and to keep OOS design compatibility in mind for future satellites. With proper support and investments in skills and capacity development, AfSA in particular has the potential to become a provider of OOS in the future. Thus, to answer the question posed by this chapter, yes, there is space for Africa in OOS, but first we need ambition, and second we need sustained support. For this reason, keeping Nkoloso and his plea (issued in 1974) in mind is timelier than ever:

> Further more, and in the next decade, we appeal to the Government to give us every assistance, materially, and morally, in order that Zambia may achieve this objective like what other nations have done in the world. I.e. to allow man, before he is dead to realise and to peep at the most wonderful glorious work of Almighty God in the outermost parts of the Universe.[83]

André Siebrits is a South African researcher focusing on the space arena (especially in developing world contexts), as well as education and the use of educational technologies, and International Relations (particularly in the Global South). He is currently working with the European Space Policy Institute (Vienna), and has experience as an e-learning researcher and as an African political risk analyst. He graduated with a Master of Arts in International Studies from the University of Stellenbosch, where his research revolved around theories of International Relations. He is currently a PhD Candidate at the Department of Political Studies at the University of Cape Town (UCT), where his research focuses on the role of the Global South in the space arena, especially in relation to governance, seen from an International Relations perspective. André is an author of publications in the e-learning field, and has written on the space-education ecosystem for sustainability and the role of educational technologies in Africa, on integrated space for African society (legal and policy implementation of space in African countries—specifically Algeria, Morocco,

education sector, see: André Siebrits and Valentino van de Heyde, "Towards the Sustainable Development Goals in Africa: The African Space-Education Ecosystem for Sustainability and the Role of Educational Technologies", in *Embedding Space in African Society: The United Nations Sustainable Development Goals 2030 Supported by Space Applications*, ed. Annette Froehlich (Cham, Switzerland: Springer, 2019).

[82]Samyukta Manikumar, "Why Natural Scientists Should Talk To Social Scientists: The Case For Interdisciplinary Dialogue In Astronomy", *Space in Africa*, 19 February 2020, https://africanews.space/why-natural-scientists-should-talk-to-social-scientists-the-case-for-interdisciplinary-dialogue-in-astronomy/.

[83]Serpell, "The Afronaut Archives: Reports from a Future Zambia".

Tunisia, and Zimbabwe), and on the African space arena. André has also presented lectures at the UCT SpaceLab (for their Space and Society course) on the African space arena and on the role of educational technologies in space education in Africa.

Chapter 6
The Space Race on Sustainability: Business and Legal Challenges for On-Orbit-Servicing

Claudiu Mihai Tăiatu

Abstract Some of the satellites operating in Earth's orbit could be regarded as critical infrastructure and some of them have implications for national security. States and societies are becoming more dependent on satellites, raising questions about the need to protect the satellites against deliberate actions or even accidental events creating space debris. Various legal questions arise about behavior in outer space. This chapter will highlight some of the legal challenges in regard of a possible commercial development of On-Orbit-Servicing (OOS) and will take into consideration the rendezvous and proximity operations through robotic missions. The analysis will also focus on the OOS dual-use implications. The business challenges for OOS will be analyzed, in particular how is the OOS relevant both for old and new generations of satellites. The European vision for a multipurpose mission including OOS capabilities will be presented. By doing so, it will highlight the position of ESA towards the space sustainability of space activities and will analyze the findings from the Space19+ ESA Ministerial Council in relation to Active Debris Removal and In-Orbit Servicing (ADRIOS) project. This chapter will also present the U.S. vision for OOS, in particular the Technology Demonstration Missions (TDM). Finally, the UNCOPUOS Long-term Sustainability (LTS) Guidelines in relation with OOS will be underlined. Recommendations for a business model for OOS will be provided. Overall, this chapter will emphasize the need to fully exploit the satellite systems already launched in GEO orbit and the possibility to manufacture in space using defunct satellites as materials. The novelty of this analysis is bringing together a wide range of space law and policy elements that would help the reader to understand what is at stake for OOS.

C. M. Tăiatu (✉)
European Space Policy Institute (ESPI), Vienna, Austria
e-mail: c.taiatu@yahoo.com

© The Editor(s) (if applicable) and The Author(s), under exclusive license to Springer Nature Switzerland AG 2020

A. Froehlich (ed.), *On-Orbit Servicing: Next Generation of Space Activities*, Studies in Space Policy 26, https://doi.org/10.1007/978-3-030-51559-1_6

6.1 Introduction

Space activities were described as an "engine for prosperity" and also an "engine for growth".[1] The global space economy was evaluated at nearly $400 billions from which three quarters was assigned to the satellite industry including services of Remote Sensing/Earth Observation, Telecommunications (television, radio, broadband, fixed and mobile), navigation, positioning and timing.[2] Space technology is an integral part of national and international infrastructures, benefits the society as a whole and is essential for realizing the achievement of the Sustainable Development Goals.[3] Space technology includes systems for the navigation of aircraft and ships, military maneuvers, financial transactions, the internet and telecommunications.[4] In particular, data and services via satellites can support various sectors among which the economy, migration, emergency response, climate change and management of natural resources.[5]

To further expand on the benefits of satellite technology, according to the International Telecommunication Union (ITU), 2019 marked the first full year when more than half of the world participated online in the global digital economy.[6] The satellite systems offer advantages for expanding broadband coverage and the satellite industry has proven its valuable contribution as a provider of instant infrastructure for commercial, government and emergency relief communications.[7] The satellite business is changing, with the NewSpace industry being disruptive while the market is evolving towards more flexible satellites and also less expensive. The demand for

---

[1]Parella, R.M.L., Space as Engine for Growth, p. 85, in: Froehlich, A., (Ed.), Post 2030-Agenda and the Role of Space: The UN 2030 Goals and Their Further Evolution Beyond 2030 for Sustainable Development, 2018, Studies in Space Policy, Volume 17, Springer International Publishing, https://doi.org/10.1007/978-3-319-78954-5 (accessed 30.03.2020). See also Preface by Froehlich, A.

[2]Bryce Space and Technology, 2019 State of the Satellite Industry, https://brycetech.com/reports.html, (accessed 12.03.2020).

[3]Wakimoto, T., Proactive Use of Artificial Intelligence for the Development: Space Satellite as a Key Infrastructure, pp. 1–11, in: Froehlich, A., (Ed.), Post 2030-Agenda and the Role of Space: The UN 2030 Goals and Their Further Evolution Beyond 2030 for Sustainable Development, Studies in Space Policy, Volume 17, 2018, https://doi.org/10.1007/978-3-319-78954-5, (accessed 30.03.2020).

[4]Unal, B., International Security Programme, Cyber and Space, July 2019, https://reader.chathamhouse.org/cybersecurity-nato-s-space-based-strategic-assets#, (accessed 14.03.2020).

[5]European Commission, Space Strategy for Europe, October 2016, https://ec.europa.eu/transparency/regdoc/rep/1/2016/EN/COM-2016-705-F1-EN-MAIN.PDF, (accessed 14.03.2020).

[6]International Telecommunication Union, The State of Broadband: Broadband as a Foundation for Sustainable Development, September 2019, https://www.itu.int/dms_pub/itu-s/opb/pol/S-POL-BROADBAND.20-2019-PDF-E.pdf, (accessed 14.03.2020).

[7]International Telecommunication Union, Non-geostationary satellite systems, December 2019, https://www.itu.int/en/mediacentre/backgrounders/Pages/Non-geostationary-satellite-systems.aspx, (accessed 12.03.2020);

See also: New podcast highlights how ITU works to close the digital divide, March 2020, https://news.itu.int/new-podcast-highlights-how-itu-works-to-close-the-digital-divide/, (accessed 12.03.2020).

satellite communications using satellites in the geostationary orbit (GEO) may slow-down, which means that it may be reconsidered its position as the most profitable business in commercial space but would probably still play an important role.[8]

The increased dependency on outer space activities, the stability and continuously development of space activities even during economic crisis, makes the satellite systems to be regarded as strategic assets.[9] The amount of information generated by space data makes the satellite systems to be vital for the benefit of humanity.[10] Being dependable on satellite technology means the satellite operator and its customers are concerned with protecting and developing their capacity to operate the satellites when needed and without interruption and the possibility to repair and upgrade the satellites. Unlawful control of the satellites could harm other space systems and such challenges fuels the growing tendency for defending these satellites. The growing number of satellites operating in space and the development of new space technologies could be seen as a threat. More satellites in orbit mean a bigger cybersecurity risk. As a result, the recent initiatives on the creation of Space Force and the inclusion of space "as a warfighting domain".[11] Defending the satellites is the concern of all spacefaring nations but the initiatives of the U.S. and NATO to create the space force as a separate branch of the military, leads by example.[12]

Satellites could be jammed, hacked or weaponized. The Report on "Space Threat Assessment" was recently issued by the Center for Strategic and International Studies (CSIS).[13] This report was produced in the context of space becoming more diverse, disruptive in terms of who is using space and how, more disordered due to the lack of

[8]Foust, J., The satellite industry catches a cold, October 2019, https://www.thespacereview.com/article/3807/1, (accessed 15.03.2020).

[9]Ibid. *supra* note Parella, R.M.L., Space as Engine for Growth, p. 95, in: Froehlich, A., (Ed.), Post 2030-Agenda and the Role of Space: The UN 2030 Goals and Their Further Evolution Beyond 2030 for Sustainable Development, 2018, Studies in Space Policy, Volume 17, Springer International Publishing, https://doi.org/10.1007/978-3-319-78954-5 (accessed 30.03.2020).

[10]Del Rio Vera, J., United Nations Office for Outer Space Affairs, Space Sustainability in the 21st Century, October 2017, at Clean Space Industrial Days, ESTEC, https://indico.esa.int/event/181/contributions/1384/attachments/1365/1590/01_23102017_CleanSpaceIndustryDays.pdf, (accessed 12.03.2020).

[11]North Atlantic Treaty Organization (NATO), Foreign Ministers take decisions to adapt NATO, recognize space as an operational domain, November 2019, https://www.nato.int/cps/en/natohq/news_171028.htm, (accessed 14.03.2020).

[12]Ellyatt, H., CNBC—Putin fears the US and NATO are militarizing space and Russia is right to worry, experts say, https://www.cnbc.com/2019/12/05/nato-in-space-putin-is-worried-about-the-militarization-of-space.html, (accessed 14.03.2020).

[13]Harrison, T., Johnson, K., Roberts, T.G., Way, T., Young, M., Space Threat Assessment 2020, March 2020, Center for Strategic and International Studies—CSIS, https://aerospace.csis.org/wp-content/uploads/2020/03/Harrison_SpaceThreatAssessment20_WEB_FINAL-min.pdf, (accessed 5.04.2020).

accepted norms and gaps in laws and treaties, and more dangerous due to the proliferation of counterspace capabilities.[14] The report contained the analysis of the types of space counterspace weapons, and the space counterspace capabilities of a number of countries. Also, Secure World Foundation (SWF) issued the Report on "Global Counterspace Capabilities: An Open Source Assessment".[15] This report includes publicly available information on the space counterspace activities, comparing each of the analyzed countries in relation to five types of space counterspace systems: direct-ascent; co-orbital kinetic antisatellite systems; electronic warfare; directed energy and cyber. Both these reports underline the reliance on satellite activities by national security and raises awareness about the development of capabilities that could interfere with the functioning of satellite systems in outer space.

In the context of the growing concerns about the integrity of the satellite systems, the question about the future deployment of On-Orbit Servicing (OOS) systems stands within the national security and military strategy for space activities. In particular, the dual-use nature of OOS raises questions about moving weapons into outer space and remains one of the main political barriers for OOS as commercial activity.[16]

The prospect of the U.S. decreasing its support or even losing its leadership role in the fight against orbital debris could raise important concerns about space sustainability.[17] It may be posed the question if there is really a decreased interest on space sustainability in general or this position favors only the NewSpace activities in LEO by facilitating the deployment of large constellations of satellites. Would such position include only the activities of Active Debris Removal (ADR) or also the OOS? In the scenario described above regarding the U.S. leadership in the fight against orbital debris, should this situation determine other states to lose interest or this is an opportunity for others to take the lead?

This research intends to present the business and legal aspects for the OOS activity, while placing the OOS analysis in an international context. Some aspects of Rendezvous and Proximity Operations (RPO) will be considered from various perspectives.

---

[14]Smith, M., Growth of Space Threats detailed in Two New Reports, March 2020, Space Policy Online, https://spacepolicyonline.com/news/growth-of-space-threats-detailed-in-two-new-reports/, (accessed 5.04.2020).

[15]Weeden, B., Samson, V., Global Counterspace Capabilities: An Open Source Assessment, April 2020, Secure world Foundation, https://swfound.org/media/206955/swf_global_counterspace_april2020.pdf, (accessed 5.04.2020).

[16]Froehlich, A., The Right to (Anticipatory) Self-Defence in Outer Space to Reduce Space Debris, p. 81, in: Froehlich, A., (Ed.), Space Security and Legal Aspects of Active Debris Removal, 2019, Studies in Space Policy, Volume 16, Springer Nature Switzerland AG, https://doi.org/10.1007/978-3-319-90338-5, (accessed 02.04.2020).

[17]Weeden, B., The United States is losing its leadership role in the fight against orbital debris, https://www.thespacereview.com/article/3889/1, (accessed 16.03.2020).

6.2 Space Sustainability

According to the Guidelines for the Long-term Sustainability (LTS) of Outer Space Activities adopted by the United Nations (UN) Committee on the Peaceful Uses of Outer Space (COPUOS) in 2019, in-orbit activities were identified as factors for affecting space sustainability:

> The proliferation of space debris, the increasing complexity of space operations, the emergence of large constellations and the increased risks of collision and interference with the operation of space objects may affect the long-term sustainability of space activities.[18]

Space sustainability is tied with the behavior in outer space. By responsible conducting maneuvers in space, the future generations would have the means to access outer space for peaceful purposes. In this context, the long-term sustainability of outer space activities was defined by UNCOPUOS as:

> the ability to maintain the conduct of space activities indefinitely into the future in a manner that realizes the objectives of equitable access to the benefits of the exploration and use of outer space for peaceful purposes, in order to meet the needs of the present generations while preserving the outer space environment for future generations.[19]

The OOS would need a system of governance for outer space behavior, the so called "rules of the road" in order to improve the safety of operations in outer space during proximity and docking operations. The space sustainability was defined as the ability to:

> ensuring that all humanity can continue to use outer space for peaceful purposes and socioeconomic benefit now and in the long term. This will require international cooperation, discussion, and agreements designed to ensure that outer space is safe, secure and peaceful.[20]

The absence of a regulatory framework for Space Traffic Management (STM) and for Space Debris Removal could negatively impact the number of collisions.[21] It will become ineffective if only a limited number of states would adopt regulations against space debris because adhering to space debris mitigation requirements would create a competitive disadvantage.[22] Within this interpretation, the measures aiming to prevent, reduce and mitigate the creation of space debris need international consensus. The space commercial activities require international cooperation

---

[18] A/AC.105/C.1/L.366, Guidelines for the Long-term Sustainability of Outer Space Activities, July 2018, https://undocs.org/A/AC.105/C.1/L.366, (accessed 15.03.2020).

[19] Ibid. *supra* note A/AC.105/C.1/L.366, Guidelines for the Long-term Sustainability of Outer Space Activities, July 2018, https://undocs.org/A/AC.105/C.1/L.366, (accessed 15.03.2020).

[20] Secure World Foundation, Space Sustainability, A Practical Guide, p. 4, 2018, https://swfound.org/media/206407/swf_space_sustainability_booklet_2018_web.pdf, (accessed 5.04.2020).

[21] Ibid. *supra* note Degrange, V., Active Debris Removal: A Joint Task and Obligation to Cooperate for the Benefit of Mankind, p. 3, in: Froehlich, A., Space Security and Legal Aspects of Active Debris Removal, 2019, Studies in Space Policy, Volume 16, Springer Nature Switzerland.

[22] Ibid. *supra* note Degrange, V., Active Debris Removal: A Joint Task and Obligation to Cooperate for the Benefit of Mankind, p. 4, in: Froehlich, A., Space Security and Legal Aspects of Active Debris Removal, 2019, Studies in Space Policy, Volume 16, Springer Nature Switzerland.

between the spacefaring nations, more coordination of these activities and the harmonization of national rules and policies. An international framework would guarantee that all actors of the space sector, including governments and private entities, would contribute to "the protection of the space environment as well as the safety of space objects".[23]

Space debris represents the highest level of concern in the space community. One of the reasons is that space debris could undermine the continued access to space. ADR and OOS are regarded as a solution to cope with the issue of space debris. In particular, OOS aims to develop robotic satellite servicing and assembly technologies.[24] It was expressed the opinion according to which the ADR operations would be more effective if not bounded by any "fault-based" legal regime according to the Liability Convention. Instead, such activity could be addressed with regard to the protection of the outer space environment and the freedom of access to outer space.[25] Even if such an interpretation could be stimulating for the ADR/OOS operators, it is challenging from a regulatory point of view. The adoption of such legal solution could have a negative impact on space sustainability as it could contribute to the growth of space debris.

The concept of space sustainability in outer space is similar with the concept of environmental protection. For this reason and to prevent a conflict, the removal of space debris could address space sustainability. In the short term, controlling the space debris will benefit the satellite operators concerned about of providing services without interruption. In the long term, a sustainable use of outer space by current generations will benefit the future generations to access space.[26] For achieving space sustainability, it would be recommended that States have initiatives aimed at facilitating transparency and confidence-building in outer space activities, mainly because of the challenges posed by ADR and OOS operations, as both are based on RPO.[27] Furthermore, in case of conflict, the international rules on the peaceful settlement of disputes could become useful in outer space activities, taking as example

---

[23]Ibid. *supra* note Degrange, V., Active Debris Removal: A Joint Task and Obligation to Cooperate for the Benefit of Mankind, p. 5, in: Froehlich, A., Space Security and Legal Aspects of Active Debris Removal, 2019, Studies in Space Policy, Volume 16, Springer Nature Switzerland.

[24]Frigoli, M., Between Active Debris Removal and Space-Based Weapons: A Comprehensive Legal Approach, p. 52, in: Froehlich, A., Space Security and Legal Aspects of Active Debris Removal, 2019, Studies in Space Policy, Volume 16, Springer Nature Switzerland.

[25]Frigoli, M., Between Active Debris Removal and Space-Based Weapons: A Comprehensive Legal Approach, p. 58, in: Froehlich, A., (Ed.), Space Security and Legal Aspects of Active Debris Removal, 2019, Studies in Space Policy, Volume 16, Springer Nature Switzerland.

[26]Nardone, V., Dispute Resolution in the Context of ADR: A Public International Law Perspective, p. 18, in: Froehlich, A., Space Security and Legal Aspects of Active Debris Removal, 2019, Studies in Space Policy, Volume 16, Springer Nature Switzerland.

[27]Nardone, V., Dispute Resolution in the Context of ADR: A Public International Law Perspective, p. 20, in: Froehlich, A., Space Security and Legal Aspects of Active Debris Removal, 2019, Studies in Space Policy, Volume 16, Springer Nature Switzerland.

the provisions of Article 33 of the UN Charter who provides the mechanism for international matters.[28]

When comparing climate change with ADR and OOS, it could be highlighted that the problems are similar, thus it is more a political decision than technical.[29] For this reason, "more pervasive dialogues and cooperation at both international and national level" are crucial for a new space policy at international level.[30] Prosperity from space activities could refer both to the activities on the ground but also to orbital activities where space commercial activities could peacefully develop if focused on the sustainable development. In this regard, the OOS could become a necessary capability for sustainable space. It is possible for instruments inside the satellite to continue to function even when the satellites will be decommissioned for the lack of fuel. In the long term, such behavior was regarded as unsustainable, contributing to the accumulation of space objects in orbit. The possibility of refueling, repairing and updating the satellites in outer space as part of OOS services is an opportunity through which the satellite operators could militate against the propagation of orbital debris.[31]

The contemporary space exploration has a commercial nature, with more private actors involved. The current situation is different from the cold war space race between the U.S. and the former Soviet Union (USSR). China, India, Japan and the Member States of the European Space Agency (ESA) have joined the U.S. and Russia as major spacefaring nations. With more actors and more private companies taking advantage of the benefits of space activities, space is becoming contested, congested and competitive. As a consequence, a new international instrument such as an international treaty on space debris removal was requested. Also, the creation of a dedicated international organization should be considered.[32] Within this regulatory framework, the OOS and ADR could gain prominence.

[28]Nardone, V., Dispute Resolution in the Context of ADR: A Public International Law Perspective, p. 22, in: Froehlich, A., Space Security and Legal Aspects of Active Debris Removal, 2019, Studies in Space Policy, Volume 16, Springer Nature Switzerland.

[29]Tian, Z., Proposal for an International Agreement on Active Debris Removal, p. 112, in: Froehlich, A., Space Security and Legal Aspects of Active Debris Removal, 2019, Studies in Space Policy, Volume 16, Springer Nature Switzerland.

[30]Ibid. *supra* note Parella, R.M.L., Space as Engine for Growth, p. 95, in: Froehlich, A., (Ed.), Post 2030-Agenda and the Role of Space: The UN 2030 Goals and Their Further Evolution Beyond 2030 for Sustainable Development, 2018, Studies in Space Policy, Volume 17, Springer International Publishing, https://doi.org/10.1007/978-3-319-78954-5 (accessed 30.03.2020).

[31]Aloia, V., The Sustainability of Large Satellite Constellations: Challenges for Space Law, p. 86, in: Froehlich, A., Legal Aspects Around Satellite Constellations, 2019, Studies in Space Policy, Volume 19, Springer Nature Switzerland, https://doi.org/10.1007/978-3-030-06028-2, (accessed 31.03.2020).

[32]Degrange, V., Active Debris Removal: A Joint Task and Obligation to Cooperate for the Benefit of Mankind, p. 2, in: Froehlich, A., Space Security and Legal Aspects of Active Debris Removal, 2019, Studies in Space Policy, Volume 16, Springer Nature Switzerland.

Space sustainability discussions are dealt at international level mainly at UNCOPUOS but also before private consortiums such as the Consortium for Execution of Rendezvous and Servicing Operations (CONFERS). The space actors operating satellites in orbit are the ones with the greatest interest in maintaining clean orbits to continue pursuing their services. According with the Outer Space Treaty (OST) provisions, States shall be responsible for national space activities whether carried out by governmental or non-governmental entities. To reach space sustainability, all space actors must be involved in supporting international measures. The spacefaring nations are the main actors, while the developing states have the voting power necessary to take decisions. So, the question about space sustainability is how fast the international community needs to act to allow RPO in LEO and GEO.

6.3 Space Activities of Non-governmental Entities

Private entities have become an integral part of the activities in outer space and some states support the developing of national space legislation to address such activities.[33] The rationale motivating the states to enact national space legislation is for the non-governmental entities to adhere to the international obligations that rest upon states. The states ensure through national legislation that activities conducted by private entities adhere to international rules.[34] The main provisions for State obligations could be found in Article VI, Article VII and Article VIII OST. These principles are further detailed in the Liability Convention and the Registration Convention. States need to authorize and continually supervise the activities conducted by private entities in outer space, the launching states will remain liable for the damage caused by the space activities conducted by non-governmental entities and also states will have the obligation to register the space objects.[35]

According to the provisions of Article VI OST, states bear international responsibility for national activities in outer space, applicable both to governmental agencies and non-governmental entities. Also, according to these provisions, the "appropriate

[33] Froehlich, A., Seffiga, V., (Eds.), National Space Legislation: A Comparative and Evaluative Analysis, p. 2, 2018, Studies in Space Policy, Volume 15, Springer International Publishing, https://doi.org/10.1007/978-3-319-70431-9, (accessed 20.03.2019).

[34] Ibid. *supra* note Froehlich, A., Seffiga, V., (Eds.), National Space Legislation: A Comparative and Evaluative Analysis, p. 7, 2018, Studies in Space Policy, Volume 15, Springer International Publishing, https://doi.org/10.1007/978-3-319-70431-9, (accessed 20.03.2019).

[35] Ibid. *supra* note Froehlich, A., Seffiga, V., (Eds.), National Space Legislation: A Comparative and Evaluative Analysis, p. 8, 2018, Studies in Space Policy, Volume 15, Springer International Publishing, https://doi.org/10.1007/978-3-319-70431-9, (accessed 20.03.2019).

state" is responsible for all activities of non-governmental entities under its jurisdiction. Thus, in line with the State responsibility, States should assure the activities are carried out in conformity with the international obligations.[36]

A particularity of space law is State liability. According to Article VII OST, States are internationally liable for damage to another State party to the OST. Further elaborated in the Liability Convention, it is notorious how complex is the concept of fault according to Article VII OST when the damage occurs in outer space.[37] In case of in-orbit collisions, the provisions of Article VII OST and the Liability Convention are "unworkable".[38] Also, according to Article VIII, States have to continually supervise such activities to be performed in accordance with provisions of national and international law. The provisions of Article VIII OST grants jurisdiction and control of a space object to the State of registry and could be regarded as customary law obligation.[39] The provisions of Article VIII OST have legal implications related to the removal of space debris, mostly regarded as a "legal obstacle that needs to be addressed for ADR to become reality".[40] However, the legal aspects remain manifestly unclear if referring to OOS commercial activity. Discussions so far showed a lack of accepted set of principles, consensus, for what space industry behavior is. As a consequence, the space industry does not have widely accepted standards of responsible space operations.[41]

The legality for individual OOS missions on the model: State A—OOS servicer State A is the most common and accepted service. In comparison, the model State B—OOS servicer State A remains uncertain. The military promoted a neutral position which is not enough anymore. There are concerns that devices used for OOS could be used for military activity and such concern needs to be clarified for the international community because without the necessary prevention rules, individual cases

---

[36] Aloia, V., The Sustainability of Large Satellite Constellations: Challenges for Space Law, p. 90, in: Froehlich, A., Legal Aspects Around Satellite Constellations, p. 90, 2019, Studies in Space Policy, Volume 19, Springer Nature Switzerland, https://doi.org/10.1007/978-3-030-06028-2, (accessed 31.03.2020).

[37] Frigoli, M., Between Active Debris Removal and Space-Based Weapons: A Comprehensive Legal Approach, p. 57, in: Froehlich, A., (Ed.), Space Security and Legal Aspects of Active Debris Removal, 2019, Studies in Space Policy, Springer Nature Switzerland.

[38] Wright, E., Legal Aspects Relating to Satellite Constellations, p. 30, in: in: Froehlich, A., Legal Aspects Around Satellite Constellations, 2019, Studies in Space Policy, Volume 16, Springer Nature Switzerland, https://doi.org/10.1007/978-3-030-06028-2, (accessed 31.03.2020).

[39] Chung, C., Jurisdiction and Control Aspects of Space Debris Removal, p. 33, in: Froehlich, A., (Ed.), Space Security and Legal Aspects of Active Debris Removal, 2019, Studies in Space Policy, Volume 16, Springer Nature Switzerland.

[40] Tian, Z., Proposal for an International Agreement on Active Debris Removal, p. 120, in: Froehlich, A., (Ed.), Space Security and Legal Aspects of Active Debris Removal, 2019, Studies in Space Policy, Volume 16, Springer Nature Switzerland.

[41] Secure World Foundation, Summary Report—Workshop on Responsible Space Behavior and the Democratization of Space, February 2020, https://swfound.org/media/206953/gstc-workshop-2020_report_mar112020-1.pdf, (accessed 15.03.2020).

for OOS will remain the rule. If no state will give an example of judicious legislation in regard of OOS, the OOS activity will resume to the same dilemma while hundreds up to thousands of satellites will be launched into space without clear rules to prevent space collisions. The trend of commercialization of outer space resulted in the need to amend the regulatory and legal framework of the space treaties and related instruments. There are several challenges to accommodate and facilitate new developments within the current regulatory and legal framework for space activities.[42]

6.4 On-Orbit Servicing (OOS)

The OOS was presented as solving a number of key problems related to satellites among which: (i) Life extension for aging assets; (ii) Part failures or anomalies, and (iii) Technological obsolescence.[43] However, because the OOS is linked with maneuvers in space, there are concerns regarding Anti-Satellite weapons (ASAT). The OOS capabilities to extend the life of satellites and transform space debris in valuable resources, by recycling non-functional satellites make the OOS to have an important role in space sustainability and become a revenue stream to fund manufacturing in space. It may be reasonably to affirm that the solution for the OOS becoming a commercial activity rests with the military strategy of the spacefaring nations. Ultimately, the question that needs to be answered at political and governmental level is if the military are interested or will become involved to support space sustainability through a commercial solution based on RPO. This challenge is augmented by the fact that:

> perhaps a space battlefield cluttered with orbital debris might even be a military advantage for the United States as it can hide, maneuver, and fight in a congested environment more easily than its adversaries, although this is not anything DOD officials have stated publicly.[44]

If there is any diplomatic solution, it thus rest with this turn of events. The answer will have to be a compromise between the military security standards and the need of commercial satellite operators to clean the orbits in order to put more capacity and operate their satellites.

To fully exploit the satellite systems already launched, it would be more cost-effective to service the satellites. OOS could be also applicable to satellite systems not designed to be serviceable. With the advancements in space technology, it would

---

[42] Ibid. *supra* note Morssink, M., An Equitable and Efficient Use of Outer Space and its Resources and the Role of the UN, the IUT and States Parties, p. 6, in: Froehlich, A., Legal Aspects Around Satellite Constellations, Studies in Space Policy, Volume 19, Springer Nature Switzerland, 2019, https://doi.org/10.1007/978-3-030-06028-2, (accessed 31.03.2020).

[43] Riesbeck, L., PowerPoint—The Critical Role of Norm-Building and Collaboration in "Standardized, Safe, and Sustainable" Commercial On-Orbit Satellite Servicing (OOS), International Astronautical Congress 2019.

[44] Ibid. *supra* note Weeden, B., The United States is losing its leadership role in the fight against orbital debris, https://www.thespacereview.com/article/3889/1, (accessed 16.03.2020).

be reasonably to predict that once the demonstration missions are successful, the satellite servicing and debris removal missions could move forward and eventually become a routine mission.

6.5 Rendezvous and Proximity Operations (RPO)

The Rendezvous and Proximity Operations (RPO) is at the core of technological advancements for OOS and ADR. RPO is the ability to operate in close proximity and rendezvous with other cooperative and/or non-cooperative space assets, including satellites.[45] Several of the advanced space capabilities under development by both governments and commercial industry, rely on RPO, meaning that such technology and capabilities will only become more widespread over time.[46]

RPO was developed in the context of co-orbital anti-satellite (ASAT) program by the former Soviet Union (USSR). The soviet Rocket booster began life in the mid-1980s as part of a Soviet co-orbital ASAT program known as Naryad: "The Naryad system utilized a rocket based on the UR-100N ICBM (NATO designation SS-19 Stiletto) that was fitted with a powerful upper stage that could place one or more kill vehicles into orbits as high as 40,000 km (24,850 miles), allowing them to independently target and home in on multiple target satellites". After 1991, the Naryad system was repurposed and this was one reason for "deepening the potential perceived link between the original Naryad program and the current RPO activities".[47] Information based on potential classified intelligence indicating RPO activities part of Russian and Chinese counterspace programs is a possible explanation for the U.S. concerns for national security. The vice versa is also possible, concerns about the U.S. space capabilities analysed by China and Russia.

Also, dating back to the beginnings of the space age, the American military developed Project SAINT (for SATellite INTerceptor). This project involved "a satellite consisting of a television camera and radar mounted in the nose of an *Agena B* upper stage. After being placed in orbit, SAINT would then maneuver close to an unfriendly target satellite, photograph and analyze it, and report back all the details to the U.S. military. The U.S. Air Force wanted to also give SAINT the ability to destroy or disable the target satellite, but such efforts were blocked by the Eisenhower and Kennedy administrations and the program was eventually canceled before it became reality". Since then, the U.S. military funded technology demonstration missions

[45] L. Riesbeck, The Critical Role of Norm-Building and Collaboration in "Standardized, Safe and Sustainable" Commercial On-Orbit Satellite Servicing (OSS), 70th International Astronautical Congress (IAC) IAC-19, E3,4,5,x52226, Washington D.C., 21–25 October 2019.

[46] Weeden, B., Dancing in the dark redux: Recent Russian rendezvous and proximity operations in space, https://www.thespacereview.com/article/2839/2, (accessed 18.03.2020).

[47] Weeden, B., Dancing in the dark redux: Recent Russian rendezvous and proximity operations in space, https://www.thespacereview.com/article/2839/2, (accessed 18.03.2020).

and experiments and concepts "similar to those found on the original non-destructive SAINT program".

The U.S. demonstrated RPO national satellite capabilities in various occasions. Successful missions of NASA Skylab mission and NASA Space Transportation System (STS)—Space Shuttle demonstrated servicing operations are possible in outer space. More recent examples are the NASA and DARPA Restore-L and Robotic Refueling Mission demonstrating RPO and OOS capabilities.[48] Also, the activity of Orbital ATK to provide life-extension services to GEO satellites using the Mission Extension Vehicle (MEV) or Mission Robotic Vehicle.

Due to the technology development, the OOS is currently developed as a robotic mission. Robots in this case could be referred to as space objects. The legal question is related with the provisions of Articles VII and VIII OST. In particular, if these legal provisions are applicable to robotic missions or if modifications are needed to cover such robotic missions.[49] It should be clarified what harmful behavior is and who will be held responsible: the software company, the OOS provider who used the software or the OOS customer who contracted the OOS or ADR services.

6.6 Demonstrated Capabilities

The RPO capabilities have been demonstrated by several spacefaring nations both in LEO and GEO. The RPO activities are being analyzed by the US national security community based on two elements: the military heritage of such activities or classified intelligence for developing counterspace capabilities. The RPO capabilities create a significant amount of concern among the main space actors and the judicious question being raised is why this is happening while they are doing similar activities and using military strategy to prove their capabilities.

6.7 U.S. Vision

The accomplishment of Northrop Grumman's Mission Extension Vehicle-1 (MEV-1) docking with the Intelsat-901 communications satellite was regarded as "a historic

[48]NASA, Space Technology Mission Directorate, Satellite Servicing TDM Project Overview, https://www.nasa.gov/mission_pages/tdm/satellite-servicing.html, (accessed 18.03.2020).

[49]Baczyńska-Wilkowska, M., Outer Space Treaty During Fourth Industrial Revolution, pp. 71–72.

rendezvous". It was the first OOS between two commercial satellites.[50] This was regarded as an exciting new era for robotic missions in orbit.[51]

The success of the U.S. defense company Northrop Grumman's MEV-1 servicer docking with an Intelsat satellite demonstrated once again that who possess OOS capabilities could provide life extension services for satellites. This OOS was a premier event as the docking occurred first time in history with a satellite that was not pre-designed with docking mechanism and also, the first time when two commercial satellites have ever docked.[52] In fact, this milestone could be considered a commercial service and not just a demonstration mission, Intelsat being the first customer for MEV-1. The Intelsat satellite was nearly 19 years old and was already placed in the graveyard orbit when the docking occurred between MEV-1 and Intelsat-901 on 25 February 2020. The MEV-1 will remain attached to Intelsat-901 and will use its own thrusters to keep the satellite properly oriented in orbit.

Upcoming initiatives to extend the life of satellites include U.S. projects. The U.S. seems to focus on GEO and is developing projects such as Made in Space's Archinaut One spacecraft.[53] NASA financed the Archinaut One project with the aim of constructing two 10-m solar arrays in orbit, highlighting that "In-space robotic manufacturing and assembly are unquestionable game-changers and fundamental capabilities for future space exploration".[54] Made in Space is looking to enable remote, in-space construction of complex structure, to eliminate spacecraft volume limits imposed as cargo by rockets and avoid the risk of spacewalks by performing some tasks by robotic missions.[55]

NASA's Restore-L robotic refueling demonstration is a robotic spacecraft equipped with technologies to rendezvous with, grasp, refuel and relocate a government-owned satellite to extend its life in LEO. The aim of this mission is to prove that servicing technologies are ready and could be integrated in other NASA

[50]Howell, E., Two private satellite just docked in space in historic first for orbital servicing, https://www.space.com/private-satellites-docking-success-northrop-grumman-mev-1.html, (accessed 12.02.2020).

[51]O'Callaghan, J., Horizon The EU Research and Innovation Magazine—Docking, rendezvous and Newton's third law—the challenge of servicing satellites in space, March 2020, https://horizon-magazine.eu/article/docking-rendezvous-and-newton-s-third-law-challenge-servicing-satellites-space.html, (accessed 16.03.2020).

[52]Henry, C., Northrop Grumman's MEV-1 servicer docks with Intelsat satellite, February 2020, https://spacenews.com/northrop-grummans-mev-1-servicer-docks-with-intelsat-satellite/, (accessed 5.03.2020).

[53]NASA, Archinaut One, October 2019, https://www.nasa.gov/mission_pages/tdm/archinaut/index.html, (accessed 18.03.2020).

[54]Berger, E., NASA seeks to break the "tyranny of launch" with in-space manufacturing, July 2019, https://arstechnica.com/science/2019/07/nasas-technology-program-funds-ambitious-in-space-manufacturing-mission/, (accessed 18.03 2020).

[55]Made in Space, For Space in Space, https://madeinspace.us/capabilities-and-technology/archinaut/, (accessed 18.03.2020).

missions.[56] Maxar Technologies will be involved in the technology development for the Restore-L that will also accommodate a payload to assemble a communication antenna and manufacture a lightweight composite beam using technology developed by Tethers Unlimited.[57]

The U.S. Defense Advanced Research Projects Agency (DARPA) Robotic Servicing of Geosynchronous Satellites (RSGS) aims to demonstrate that a robotic mission could perform satellite servicing operations on operational GEO satellites and be launched with sufficient propellant and payload robustness to allow more than one mission over several years.[58] DARPA signed a partnership with a subsidiary of Northrop Grumman Corporation to become the commercial partner of RSGS program.[59]

6.7.1 Success of the OSS Demonstration Mission Brings Additional Funding

A notable consequence from the OOS demonstration in GEO is the U.S. Defense Advanced Research Projects Agency (DARPA) selecting SpaceLogistics LLC, a subsidiary of Northrop Grumman Corporation, as its commercial partner for the Robotic Servicing of Geosynchronous Satellites (RSGS) program. According to DARPA, "Space Logistics will provide the spacecraft bus based upon technologies from its Mission Extension Vehicle line, integrate the resulting robotic servicing spacecraft with the launch vehicle and provide the launch, as well as the mission operations center and staff for the full mission duration".[60] Within this program, DARPA aims to provide the first-ever commercial robotic servicing spacecraft that will perform missions both for commercial and government client satellites with advanced robotics technology, including "in-orbit repair, augmentation, assembly, detailed inspection and relocation of client satellites".[61]

---

[56] NASA, Restore-L Robotic Servicing Mission, https://sspd.gsfc.nasa.gov/restore-l.html, (accessed 18.03.2020).

[57] Writers, S., NASA funds demonstration of assembly and manufacturing in space, February 2020, https://www.spacedaily.com/reports/NASA_funds_demonstration_of_assembly_and_manufacturing_in_space_999.html, (accessed 18.03.2020).

[58] Parrish, J., DARPA—Robotic Servicing of Geosynchronous Satellites, https://www.darpa.mil/program/robotic-servicing-of-geosynchronous-satellites, (accessed 18.03.2020).

[59] Ibid. *supra* note DARPA, In-space Robotic Servicing Program Moves Forward with New Commercial Partner, March 2020, https://www.darpa.mil/news-events/2020-03-04, (accessed 18.03.2020).

[60] DARPA, In-space Robotic Servicing Program Moves Forward with New Commercial Partner, March 2020, https://www.darpa.mil/news-events/2020-03-04, (accessed 14.03.2020).

[61] SpaceLogistics selected by DARPA as commercial partner for robotic servicing mission, March 2020, https://www.controldesign.com/industrynews/2020/spacelogistics-selected-by-darpa-as-commercial-partner-for-robotic-servicing-mission/, (accessed 14.03.2020).

6.8 European Vision

Europe is focused on finding solutions to reduce space debris in LEO and to support space sustainability. During the Space19+ ESA's Ministerial Council, ESA's announced the world's first space debris removal mission, planned for launch in 2025. Space19+ Conference gave Europe a technological edge, in particular to automate collision avoidance or remove most hazardous debris. ClearSpace-1 will be the first space mission to remove an item of debris from orbit, planned for launch in 2025.[62]

ESA is Europe's gateway to space and its purpose is "to provide for and to promote, for exclusively peaceful purposes, cooperation among European States in space research and technology".[63] In this regard, ESA's Member States strongly support commercial services in the future for space sustainability. According to ESA, they will be developing essential guidance, navigation and control technologies and rendezvous and capture methods trough ADRIOS project. Its name comes from the initials of Active Debris Removal/In-Orbit Servicing. These findings from this ESA project will be applied to ClearSpace-1 mission which will allow ESA to demonstrate these technologies, in a world's premier. According to ESA, the ClearSpace-1 mission will target an upper stage "Vespa"—Vega Secondary Payload Adapter, at approximately 660 km. As a result, the chaser and the target will burn up in the atmosphere during de-orbiting.[64] The ClearSpace-1 is a concept developed by ESA at Space19+ which differs from Clean Space initiative from the 2012 ESA Ministerial Council. The ClearSpace-1 has broadened ESA's from just an ADR mission to a multi-purpose Space Servicing Vehicle that would include debris removal, refueling capabilities, upgrading or re-positioning.[65] The objective is to develop a multi-purpose In-Orbit Servicing Vehicle (IOSV) able to perform a variety of operations in orbit including in the field of active debris removal.[66]

Swiss startup ClearSpace plans to begin leading a European consortium focused on capturing a Vespa payload adapter in 2025 and dragging it into Earth's atmosphere.[67] ESA funding will cover spacecraft development and launch costs and the total mission was estimated at 117 million Euro. Luc Piguet is the co-founder and chief executive of ClearSpace and according to the news, he revealed that the first mission will destroy both the space debris and the servicer spacecraft, meaning that they will both de-orbit and burn into the atmosphere, while future plans for servicers is to de-orbit multiple

---

[62] ESPI, STM, p. 53.

[63] ESA Convention and Council Rules of Procedure, SP-1317/EN, December 2010, https://esamultimedia.esa.int/docs/LEX-L/ESA-Convention/SP-1317_EN.pdf, (accessed 18.03.2020).

[64] European Space Agency, ESA commissions world's first space debris removal, December 2019, https://www.esa.int/Safety_Security/Clean_Space/ESA_commissions_world_s_first_space_debris_removal, (accessed 16.03.2020).

[65] European Space Agency, In-Orbit Servicing: Disposal, August 2019, https://www.esa.int/Safety_Security/Clean_Space, (accessed 16.03.2020).

[66] ESPI, Space Traffic Management, p. 52.

[67] SpaceNews, 20 Space Industry Predictions for 2020, https://spacenews.com/20-space-industry-predictions-for-2020/, (accessed 14.03.2020).

objects without also to destroy themselves.[68] Such an approach will be more cost-effective. ESA developing a Space Servicing oriented mission could be seen also as financially driven as the servicing capabilities could bring the return of investment.

In contrast, the objective of ESA's Clean Space initiative and e.Deorbit debris removal mission was to de-orbit the ESA-owned Envisat satellite from LEO. The Active Debris Removal/In-Orbit Servicing was previously known as e.Deorbit. It was mentioned that the initial approach to de-orbit the target on a service-orientated basis to private companies was seen as challenging and risky.[69]

6.9 OOS Legal Challenges

The absence of a legal framework for NewSpace activities allowed the military and geopolitical interests to fill the legal gaps.[70] In particular, the legality of the removal of space debris is affected by the lack of a legal definition of the term "space debris".[71] The lack of norms to regulate behavior in space and lack of international agreed registry for satellite location in space could be regarded as the biggest challenges for OOS. The "Treaty on Principles Governing the Activities of States in the Exploration and Use of Outer Space, including the Moon and Outer Celestial Bodies" (OST) provides the basic principles for the activity in outer space.[72] Neither the OST nor the other four treaties supplementing OST are sufficient in providing a solution for the commercial activity for OOS. Demonstration missions for the OOS took place under the current framework, but these are only individual cases and were most likely supplemented by bilateral agreements between the satellite operator and OOS servicer. The OST framework is mainly built as a "treaty of principles" and "governing the activities of States". Thus, the OST does not provide legal solutions to all the scenario in a routine operational activity and for commercial purposes.

OOS is a complex space mission, that would be performed by machines, which would need a meticulous preparation and a large system of sensors and radars to feed

[68] Henry, C., Swiss startup ClearSpace wins ESA contract to deorbit Vega rocket debris, December 2019, https://spacenews.com/swiss-startup-clearspace-wins-esa-contract-to-deorbit-vega-rocket-debris/, (accessed 16.03.2020).

[69] European Space Agency, ESA's e.Deorbit debris removal mission reborn as servicing vehicle, December 2018, https://www.esa.int/Safety_Security/Clean_Space/ESA_s_e.Deorbit_debris_removal_mission_reborn_as_servicing_vehicle, (accessed 16.03.2020).

[70] Frigoli, M., Wild Military Operations in Outer Space: A Sword of Damocles Hanging over the Future of Space Environment and Space Activities, p. 51, in Froehlich, A. (Ed.), A Fresh View on the Outer Space Treaty, 2018, Studies in Space Policy, Volume 13, Springer International Publishing.

[71] Froehlich, A., The Right to (Anticipatory) Self-Defence in Outer Space to Reduce Space Debris, p. 73, in: Froehlich, A., (Ed.) Space Security and Legal Aspects of Active Debris Removal, 2019, Studies in Space Policy, Volume 16, Springer Nature Switzerland.

[72] Res 2222 (XXI), Treaty on Principles Governing the Activities of States in the Exploration and Use of Outer Space, including the Moon and Other Celestial Bodies, 1966, https://www.unoosa.org/oosa/en/ourwork/spacelaw/treaties/introouterspacetreaty.html, (accessed 18.03.2020).

the algorithms, which are currently not in place. Without Space Situational Awareness (SSA) infrastructure, large scale OOS operations would not be possible to be performed on a commercial basis. For these reasons, the current space law framework needs to be developed to support a sustainable commercial activity. Until then, the legal solution could be given by contractual stipulations. Modern space operational domains were identified as scientific, commercial, intelligence and military and highlighted that these are governed by international and national law, governmental and corporate policies, customs and precedents, and contracts.[73]

Legal challenges for OOS were identified in regard of liability. In particular, there are limitations in monitoring space activities, there are no rules of the road and therefore there are challenges of proving fault. In such case, challenges about liability would be necessary to be dealt through bilateral agreements.[74] Related to the liability concerns, there could be even stipulated a cross-waiver of liability between the satellite operator and the OOS provider from the same country, the risk being able to be supported through insurance. The questions persisting would have to be dealt between the parties. If a service provider from country A could perform an OOS mission for country B, in case of an incident who would be responsible: Country A, Country B, the satellite operator, or the OOS provider? The space law framework should provide adequate legal solutions at such kind of questions. Also, the space law framework for OOS should provide an effective procedural means to manage a collision that may happen in orbit during servicing operations. On-Orbit Operations will become a routine and it may be reasonable to seek legal solutions to manage a collision without the risk of conflict in space. Such a solution should take into consideration not to risk the OOS provider business which in most cases will lack the funding to cover the liability. If a collision will happen, the most probably the costs for the insurance and launching will raise. The success of the OOS demonstration missions should support the work of the legislators and should become a strong argument for international cooperation. Planning ahead will avoid the risk of conflict in outer space in case of a collision.

The provisions of Article VIII OST grant jurisdiction and control for an indeterminate period of time, resulting that a state is entitled to exercise its sovereignty over its registered space objects.[75] According to Article VIII OST provisions, the registering State of a space object "shall retain jurisdiction and control" while in outer space or even on a celestial body. The jurisdiction and control applies to space debris and any action over the space objects would not be possible without the consent of

---

[73]Ibid. *supra note* Riesbeck, L., PowerPoint—The Critical Role of Norm-Building and Collaboration in "Standardized, Safe, and Sustainable" Commercial On-Orbit Satellite Servicing (OOS), International Astronautical Congress 2019.

[74]Dethlefsen, T.F., On-Orbit Servicing: Repairing, Refueling and Recycling the Legal Framework, IAC-19-E7,1,7,x53939, https://iislweb.org/wp-content/uploads/2019/11/IAC-2019_Thea_F_Dethlefsen.pdf, (accessed 17.03.2020).

[75]Chung, C., Jurisdiction and Control Aspects of Space Debris Removal, p. 33, in: Froehlich, A., Space Security and Legal Aspects of Active Debris Removal, 2019, Studies in Space Policy, Volume 16, Springer Nature Switzerland.

the registration state.[76] If a removal operation would be performed unilaterally by a third State, it would constitute a violation of the OST.[77] When referred as a double-sword, the jurisdiction and control clause means in practice that "while other nations are prohibited from removing registered space objects of another state, the State of registry – as the appropriate state – is liable for any damage caused by the debris of such space objects".[78] The legitimacy of removing a registered space object of another state, even in case of posing a danger, was not given a legal solution. If the jurisdictional power granted by Article VIII is permanently attached to the State of registry or it would be applicable only during the physical control.[79] It was expressed that the "exact nature" of the registering procedure and powers attributed therefore to the registration states is a pre-condition to facilitating the development of space debris removal.[80] The questions raised by the provisions of Article VIII OST are similar for the OOS as they are for ADR. This will apply in the case the OOS provider would like to use the defunct satellites as spare parts.

Also, the upgrades of the satellites, or the objects resulted from recycling the satellites should be dealt under the provisions of Article VIII OST. Related to the transfer of ownership of the satellites, in absence of an agreement, the new owner would have to take into consideration that the registering state would retain jurisdiction and control over the upgrades of the satellite. Such obligations are of concern also in the case of on-orbit recycling, when a new owner and a new object would exist. Should it have the consent, the new owner would need to register in accordance with Registration Convention and the launching State description.

The unwillingness to amend or create new treaties has revealed the opportunity to establish soft law, such as the Space Debris Mitigation Guidelines. Debris mitigation is addressed by the current guidelines while the debris remediation is not.[81] Alternatively, it was expressed the need of a new set of guidelines that will address

[76]Chung, C., Jurisdiction and Control Aspects of Space Debris Removal, p. 40, in: Froehlich, A., Space Security and Legal Aspects of Active Debris Removal, 2019, Studies in Space Policy, Volume 16, Springer Nature Switzerland.

[77]Frigoli, M., Between Active Debris Removal and Space-Based Weapons: A Comprehensive Legal Approach, p. 56, in: Froehlich, A., Space Security and Legal Aspects of Active Debris Removal, 2019, Studies in Space Policy, Volume 16, Springer Nature Switzerland.

[78]Chung, C., Jurisdiction and Control Aspects of Space Debris Removal, p. 41, in: Froehlich, A., Space Security and Legal Aspects of Active Debris Removal, 2019, Studies in Space Policy, Volume 16, Springer Nature Switzerland.

[79]Chung, C., Jurisdiction and Control Aspects of Space Debris Removal, p. 46, in: Froehlich, A., Space Security and Legal Aspects of Active Debris Removal, 2019, Studies in Space Policy, Volume 16, Springer Nature Switzerland.

[80]Chung, C., Jurisdiction and Control Aspects of Space Debris Removal, p. 45, in: Froehlich, A., Space Security and Legal Aspects of Active Debris Removal, 2019, Studies in Space Policy, Volume 16, Springer Nature Switzerland.

[81]Tian, Z., Proposal for an International Agreement on Active Debris Removal, p. 110, in: Froehlich, A., Space Security and Legal Aspects of Active Debris Removal, 2019, Studies in Space Policy, Volume 16, Springer Nature Switzerland.

space debris removal.[82] The OOS and ADR could replace the legal vacuum and create more certainty for the private operators seeking profit coming from ADR/OOS space commercial activities.

As an alternative to the legal framework, some aspects for OOS could be dealt by:

Non-legally binding instruments in the form of Guidelines for OOS operations and design to support safe operations and increase transparency;

Implement these guidelines into national licenses for OOS operations to make them legally binding;

Contractually allocate risks related to liability, export control, insurance requirements.[83]

Ideally, States would agree on a new international treaty addressing space debris, including OOS and ADR services. Article IX OST regarding due regard to the corresponding interests of all State Parties could be a good starting point for a new treaty on space sustainability. However, since states are currently reluctant to enact new binding legal instruments, other solutions should be found in the near future. The major trend both for ADR and OOS is to voluntarily adopt guidelines and standards, such as for de-orbiting, maneuvering in space and cybersecurity and the governments seem to support such private initiatives. The goal is to help building transparency and confidence between space faring nations. A solution mainly for the ADR services would be including debris removal clauses in national licensing requirements. Such clauses could *inter alia* require the licensee to produce an insurance to cover the costs of removal of the satellite in case the planned disposal procedures at end of life are impeded.

To become legally binding, the OOS guidelines should be further implemented by national authorities. In such case, the requirement would be applicable to license applications for OOS activities.

The lack of norms is both an opportunity and a challenge for space industry players. The opportunity comes from the possibility of the commercial players to develop a practice that would benefit the sustainable development of their business model. This is the example of the Consortium for Execution of Rendezvous and Servicing Operations (CONFERS) which "aims to leverage best practices from government and industry to research, develop, and publish non-binding, consensus-derived technical and operations standards" for OOS and RPO.[84] Also, Space Safety

[82] De Waal Alberts, A., The Degree of the Lack of Regulation of Space Debris Within the Current Space Law Regime and Suggestions for a Prospective Legal Framework and Technological Interventions, p. 105, in: Froehlich, A., Space Security and Legal Aspects of Active Debris Removal, 2019, Studies in Space Policy, Volume 16, Springer Nature Switzerland.

[83] Ibid. *supra* note Dethlefsen, T.F., On-Orbit Servicing: Repairing, Refueling and Recycling the Legal Framework, IAC-19-E7,1,7,x53939, https://iislweb.org/wp-content/uploads/2019/11/IAC-2019_Thea_F_Dethlefsen.pdf, (accessed 17.03.2020).

[84] Barnhart, D.A., Rughani, R., On-Orbit Servicing Ontology applied to Recommended Standards for Satellites in Earth Orbit, IAC-19-D1.6.9, https://www.isi.edu/sites/default/files/users/barnhart/IAC-19-D1.6.9-On-Orbit%20Servicing%20Ontology%20applied%20to%20Recommended%20Standards%20for%20Satellites%20in%20Earth%20Orbit-FINAL.pdf, (accessed 3.03.2020).

Coalition (SSC) gathers international space actors, such as satellite operators, satellite manufacturers, investors, customers, policy makers, military to be part of a multilateral dialogue for securing space sustainability.[85] The developing of standards and norms as best practices, benefits the space industry players but also policy makers that look for international consensus.

The lack of norms and legal gaps is also an important challenge for space sustainability because the RPO maneuvers could be categorized as harmful behavior. Thus, the lack of behavior norms in outer space or the so-called "rules of the road" or Space Traffic Management (STM) is a challenge for commercial OOS. The dual-use applications make the transparency and confidence building measures (TCBM) an essential tool in the advancement of On-Orbit Servicing and Active Debris Removal. International cooperation is needed to provide sufficient guarantees that would prevent military conflicts and would develop norms of behavior and space situational awareness (SSA). Also, there is a need of transparency which would allow the international space actors to develop their communication channel in order to prevent conflict and/or decrease risk of collision.

For commercial actors, the norms are developed by commercial practice, tested, refined and eventually adapted into legal systems.[86] Norms of operation are necessary for the oversight of RPO, OOS and ADR capabilities and essential to identify abnormal behavior and investigation. Currently, no international nor national treaty regime governs OOS activities. Such activity is not forbidden, but the legal challenges hinders this technology to be commercially exploited. A number of space companies already have contracts with satellite operators for servicing missions, but such contracts are not public, they are not standardized. A very complex systems make the information to be corroborated from international treaties and principles, domestic legislation, government and corporate policies, customs and precedents and commercial contracts.[87] The OOS should take into consideration the necessary steps to achieve a "standardized, safe and sustainable" activity. To achieve the ideal situation poses a variety of policy questions and challenges.

A U.S. satellite or a satellite of another country that has U.S. components or technology on board falls within the definition of "export" under the ITARs. Thus, in the absence of any modification to the United States International Traffic in Arms Regulations (ITARs), the ADR and OOS orbital services for U.S. space objects and

---

[85]Space Safety Coalition, Best Practices for the Sustainability of Space Operations, September 2019, https://spacesafety.org/best-practices/, (accessed 17.03.2020).

[86]Riesbeck, L., The Critical Role of Norm-Building and Collaboration in "Standardized, Safe and Sustainable" Commercial On-Orbit Satellite Servicing (OOS), IAC-19, E3,4,5,x52226, https://iafastro.directory/iac/proceedings/IAC-19/IAC-19/E3/4/manuscripts/IAC-19,E3,4,5,x52226.pdf, (accessed 18.03.2020).

[87]Ibid. *supra* note Riesbeck, L., The Critical Role of Norm-Building and Collaboration in "Standardized, Safe and Sustainable" Commercial On-Orbit Satellite Servicing (OOS), IAC-19, E3,4,5,x52226, https://iafastro.directory/iac/proceedings/IAC-19/IAC-19/E3/4/manuscripts/IAC-19,E3,4,5,x52226.pdf, (accessed 18.03.2020).

not only, should be provided by space companies with the same nationality with the registration state.[88]

The legal framework also needs to address the implication about the dual-use technologies of OOS missions. The UN could arbitrate such a situation and prevent tensions in outer space related to security, assure the transparency of operations and registration. The implementation of an international STM and SSA would be desirable in order to assure safety of operations and security of space assets. A legal framework should take into consideration technical and regulatory safety standards.

6.10 Leadership in Space Sustainability—OOS Perspective

Do we have leaders in space activities? The short answer is yes and is based on strong proof of international investment in space activities and development of technological outbreaks that support the development of the human race as a whole. Space Law and Policy is essential to shape governance in space activities and has been successful in many occasions. Is there a leader in the domain of space debris mitigation? We can definitely talk about different initiatives in support of space debris mitigation, both at international and national level, private and public, government, industry and academia. So far there is an incipient stage, with the space industry trying to figure out a business model, in order to make the space debris mitigation activity profitable.

The US Space Policy Directive-3 is an example of how the commercial leadership could be integrated with the international obligations for space sustainability. One of the objectives the US Space Policy Directive-3 National Space Traffic Management Policy is to "encourage and facilitate U.S. commercial leadership in Science and Technology, Space Situational Awareness, and Space Traffic Management".[89] The policy aims "to guarantee favorable safety and regulatory conditions for the emergence and growth of new commercial space ventures and activities involving, for example, on-orbit servicing, debris removal, in-space manufacturing, space tourism, small satellites or very large constellations".[90]

---

[88] Tian, Z., Proposal for an International Agreement on Active Debris Removal, p. 110, in: Froehlich, A., Space Security and Legal Aspects of Active Debris Removal, 2019, Studies in Space Policy, Volume 16, Springer Nature Switzerland.

[89] Space Policy Directive-3, National Space Traffic Management Policy, June 2018, https://www.whitehouse.gov/presidential-actions/space-policy-directive-3-national-space-traffic-management-policy/, (accessed 18.03.2020).

[90] European Space Policy Institute, Towards a European Approach to Space Traffic Management, p. 32, in: ESPI Report 71, January 2020, https://espi.or.at/publications/espi-public-reports, (accessed 17.03.2020).

6.11 Dual—Use Implications

The Outer Space has been described as a militarized environment. A reason for explaining why the military will always underline the risk of conflict in space is because the original considerations for entering space activities are military. The militarization of outer space and the weaponization of outer space are different concepts, with the latter being related to the dual-use nature of OOS. The weaponization of space refers to antisatellite weapons, for example the deployment of satellites with offensive capabilities against a target located in space.[91] Antisatellite weapons, in particular the co-orbital space objects, are particularly relevant in discussing the risk of OOS services. The OOS space object is intended to remain in orbit reason for which it could be considered dangerous. Attention to ASATs is attributed to major spacefaring nations and it is considered only a matter of time before other spacefaring nations will developed such capabilities.[92]

If referring to Article IV OST, these provisions represent a restriction of placing in orbit around the Earth, objects carrying nuclear weapons or any other kinds of weapons of mass destruction, install such weapons on celestial bodies, or station such weapons in outer space in any other manner. Interpreting these provisions, the placement in outer space in orbits around the Earth of antisatellite weapons is not forbidden.[93] The military uses of outer space were described as heading to space weaponization, reason for which some states could be reluctant to the OOS activities.

Weapon systems other than nuclear weapons and/or weapons of mass destruction are not subject to the restriction mentioned in Article VI, paragraph 1 OST. A text that would have forbidden any kind of weapon into Earth orbit or placed on celestial bodies does not exist. An attempt was made with the "Treaty on Prevention of the Placement of Weapons in Outer Space and the Threat or Use of Force against Outer Space Objects" (PPWT) proposed by China and Russia. But this instrument met strong opposition and is not in force.[94] Interpreting the provisions of Article IV paragraph 4 OST, the use of "military personnel for scientific research or for any

[91] Frigoli, M., Wild Military Operations in Outer Space: A Sword of Damocles Hanging over the Future of Space Environment and Space Activities, p. 51, in: Froehlich, A., (Ed.), A Fresh View on the Outer Space Treaty, 2018, Studies in Space Policy, Volume 13, Springer International Publishing.

[92] Frigoli, M., Wild Military Operations in Outer Space: A Sword of Damocles Hanging over the Future of Space Environment and Space Activities, p. 54, in: Froehlich, A., (Ed.), A Fresh View on the Outer Space Treaty, 2018, Studies in Space Policy, Volume 13, Springer International Publishing.

[93] Frigoli, M., Wild Military Operations in Outer Space: A Sword of Damocles Hanging over the Future of Space Environment and Space Activities, p. 55, in: Froehlich, A., (Ed.), A Fresh View on the Outer Space Treaty, 2018, Studies in Space Policy, Volume 13, Springer International Publishing.

[94] Froehlich, A., The Right to (Anticipatory) Self-Defence in Outer Space to Reduce Space Debris, p. 81, in: Froehlich, A., (Ed.) Space Security and Legal Aspects of Active Debris Removal, 2019, Studies in Space Policy, Volume 16, Springer Nature Switzerland.

other peaceful purposes" means that the military personnel would be allowed in the context of space debris removal.[95]

To allow such activities, more states would have to accept the OOS services in support of the peaceful exploration and use of outer space. The provisions of Article XI OST related to obligations of "no harmful contamination of outer space" and of "no harmful interference with outer space activities of other state-parties" could be regarded as an offset to the military activities in outer space. "With the state practice directed toward the threshold of weaponization and a growing population of space debris, Article IX OST should be considered the core of the future regulation of outer space activities".[96]

The RPO technology is dual use, meaning that it could benefit both civilian and military purposes, and therefore the development of OOS and ADR could have implications on both the military and space private/commercial companies. OOS is a good example where discussions include both the commercial and military activity and where is a strong need for international cooperation between Governments, Industry and Academia. The development and the future use of OOS technology is multipurpose and can benefit space sustainability, reason for which the business and legal considerations should be analyzed from manifold perspectives.

OOS capabilities raise important concerns for the military, especially related to national security, as "most of the technologies and capabilities that provide OOS could also be used to intentionally harm satellites and could be considered offensive counterspace capabilities".[97] Historically, RPO technology was developed by military. In this context, the transparency and confidence building measures (TCBM) play an essential role in the advancement of OOS and ADR as commercial activity. The new services provided by OOS space companies could not be delivered by commercial entities without international cooperation, in particular without military observation, because the lack of the government involvement, at least as an observer, would drastically increase the risk of conflict. There are multiple objectives of the servicing mission. However, OOS will always maneuver in close proximity of the target and will rendezvous with it. The main concern is that the OOS missions will always approach satellites.

[95] Ibid. *supra* note Froehlich, A., The Right to (Anticipatory) Self-Defence in Outer Space to Reduce Space Debris, p. 81, in: Froehlich, A. (Ed.), Space Security and Legal Aspects of Active Debris Removal, 2019, Studies in Space Policy, Volume 16, Springer Nature Switzerland.

[96] Frigoli, M., Wild Military Operations in Outer Space: A Sword of Damocles Hanging over the Future of Space Environment and Space Activities, p. 57, in: Froehlich, A., (Ed.), A Fresh View on the Outer Space Treaty, 2018, Studies in Space Policy, Volume 13, Springer International Publishing.

[97] Weeden, B., Intel…Zombiesats and On-Orbit Servicing, September 2010, https://www.milsatmagazine.com/story.php?number=152118614, (accessed 3.3.2020).

The dual use was plastically referred to as: "one man's ADR system is another man's ASAT".[98] The capabilities of towing satellites in space would be referred to as a dangerous tool used to tow enemy satellites to their demise and the commercial activity such could the ADR be could be considered a cover for an illegal space-based ASAT weapon program.[99] Because of the dual-use nature, the ADR was regarded as a delicate topic for international policy: "ADR technologies are a delicate topic in international space politics, since any such system could be used against a rival nation's space assets. The dual-use nature of all space hardware and ADR in particular means that any technical solution must be at the behest of the international space community at large and not of one single nation or agency".[100] The OOS could also have the same issue. The International Space Station (ISS) was given as an example for spacefaring nations working collaboratively with dual-use technology without compromising national security.[101] The ISS is the best example of international cooperation in space activities were space remains peaceful and supports cooperative partners aboard the ISS. The 1998 IGA is not compromising sensitive information.

6.12 OOS Operations

The analysis of the OOS and ADR was made in various scientific papers, including technical and legal perspective. The OOS and ADR were proposed as international solutions for mitigating space debris.

The RPO technology is similar for OOS and ADR, and multipurpose missions are currently being developed. But the problem however is the difference between LEO

[98] Wilhelm, J.C., The Keys to Rule Them All: Sustainable Development of Orbital Resources, p. 66, in: Froehlich, A., (Ed.), Post 2030-Agenda and the Role of Space: The UN 2030 Goals and Their Further Evolution Beyond 2030 for Sustainable Development, 2018, Studies in Space Policy, Volume 17, Springer International Publishing, https://doi.org/10.1007/978-3-319-78954-5 (accessed 30.03.2020).

[99] Ibid. *supra* note Wilhelm, J.C., The Keys to Rule Them All: Sustainable Development of Orbital Resources, p. 67, in: Froehlich, A., (Ed.), Post 2030-Agenda and the Role of Space: The UN 2030 Goals and Their Further Evolution Beyond 2030 for Sustainable Development, 2018, Studies in Space Policy, Volume 17, Springer International Publishing, https://doi.org/10.1007/978-3-319-789 54-5 (accessed 30.03.2020).

[100] Ibid. *supra* note Wilhelm, J.C., The Keys to Rule Them All: Sustainable Development of Orbital Resources, p. 67, in: Froehlich, A., (Ed.), Post 2030-Agenda and the Role of Space: The UN 2030 Goals and Their Further Evolution Beyond 2030 for Sustainable Development, 2018, Studies in Space Policy, Volume 17, Springer International Publishing, https://doi.org/10.1007/978-3-319-789 54-5 (accessed 30.03.2020).

[101] Ibid. *supra* note Wilhelm, J.C., The Keys to Rule Them All: Sustainable Development of Orbital Resources, p. 69, in: Froehlich, A., (Ed.), Post 2030-Agenda and the Role of Space: The UN 2030 Goals and Their Further Evolution Beyond 2030 for Sustainable Development, 2018, Studies in Space Policy, Volume 17, Springer International Publishing, https://doi.org/10.1007/978-3-319-789 54-5 (accessed 30.03.2020).

Table 6.1 Comparative presentation of space activities in LEO and GEO

LEO	ADR	Cheap satellites, easily replaceable	5 years of life	Cost are not recoverable from satellite services, but support sustainability
GEO	OOS	Very expensive satellites, expensive to be replaced	15 years of life	Cost are recoverable from satellite services revenue, and support sustainability

and GEO and the cost of the satellites being deployed. In LEO, the trend of mega constellations proposes cheap satellites easily replaceable. Servicing of the cheap satellites in LEO makes no economic sense, being preferably sending a new satellite to replace the old one (see Table 6.1).

According to the above-mentioned data, ADR does not make sense for GEO and OOS does not make sense for LEO. Taking the as an example ADRIOS, ESA project for ADR and OOS, a spacecraft being able to do the both missions would make sense to develop.

The business case for OOS implies an economic return from the mission. Democratization of space indicates that space activities are expanding to a growing number of states and non-state actors.[102] Governments are the main investors in space activities, via procurement and grants mechanism to public agencies, research institutes, universities and the private sector, and the returns from investments in space programmes are of a diverse nature.[103] While the private-sector companies focus on developing space technologies, the return of investment, at least in space exploration, is mainly from public funding. The public private partnerships role to transfer the risk, money spending and time from exclusively governmental source to a private enterprise could raise the question about a commercial space race or government risk sharing. For example, the United States have the biggest government budget for space activities, suppressing the combined Non-US Government Budgets by half.[104] Another example is ESA's Council at Ministerial Level, Space19+ where ESA ministers committed to the ever-biggest budget with total subscriptions of more than EUR 14 billion.[105]

[102]Pekkanen, S.M., Governing the New Space Race, April 2019, in: American Journal of International Law, AJIL Unbound, Volume 113, 2019, pp. 92–97, https://www.cambridge.org/core/journals/american-journal-of-international-law/article/governing-the-new-space-race/14BD9B37A7A15A8E225A5355BB29E51B/core-reader#, (accessed 15.03.2020).

[103]Organisation for Economic Co-operation and Development (OECD), The Space Economy in Figures: How Space Contributes to the Global Economy, July 2019, https://www.oecd.org/innovation/the-space-economy-in-figures-c5996201-en.htm, (accessed 12.03.2020).

[104]Bryce Space and Technology, 2018 Global Space Economy, 2018, https://brycetech.com/reports.html, (accessed 12.03.2020).

[105]European Space Agency, No. 22—2019: ESA ministers commit to biggest ever budget, November 2019, https://www.esa.int/Newsroom/Press_Releases/ESA_ministers_commit_to_biggest_ever_budget, (accessed 14.03.2020).

6.13 Manufacturing in Space

Manufacturing in space is a promising business. In many processes, it was concluded that microgravity is preferable over normal gravity and this is a valuable opportunity for commercial players to take the lead. Examples are many in public space. It was highlighted that microgravity culture enhances the therapeutic effect of stem cells transplantation.[106] The Microgravity Cell Biology is important both as a commercial business opportunity but also scientific as it could increase our understanding of the role of gravity in life processes.[107] The fluoride glass optical fiber is another example of commercial manufacturing in space. Commonly known as ZBLAN, the fluoride glass optical fiber if manufactured in space would become far more valuable than on Earth as it could avoid the defects that occur during the process of solidification.[108]

The ability to manufacture in space is critical for space exploration, especially for building facilities in other regions of space such as planets. Materials already in space could become useful for the process of 3D printing.

6.13.1 OOS Recycling Business

The most obvious benefit of OOS is to increase the lifetime of satellite systems through refueling, upgrade or repositioning. This service could diminish the need of launching new satellites in GEO to replace the old ones. However, this is only one use of the OOS technology, with more commercial opportunities of such technology. The full spectrum of opportunities is broader, and automation is the key for manufacturing large structure in orbit with robots in space. For example, turning Space Debris into a resource is not enough to have the capability to dock and remove the satellite, but you need servicing capabilities that you can perform in orbit. Opportunities to recycle abound, even in outer space.

It may be posed the question is if there are any legal constraints for delivering OOS services in space. The servicing of satellites in outer space is not forbidden. What is challenging is that under the current legal framework, not all the satellites could be serviced without special authorizations. Several commercial companies are investing in developing OOS capabilities and contracts have been signed with customers from governments and private sector. Among the most obvious applications is refueling missions. Refueling missions to satellites in GEO is a right-away profitable option

---

[106] Imura, T., Otsuka, T., Kawahara, Y., Yuge, L., "Microgravity" as a unique and useful stem cell culture environment for cell-based therapy, 15 December 2019, Regenerative Therapy, Volume 12, pp. 2–5, https://doi.org/10.1016/j.reth.2019.03.001, https://www.sciencedirect.com/science/article/pii/S2352320418300956, (accessed 5.03 2020).

[107] Pellis, N.R., Microgravity Cell Biology, https://www.nasa.gov/pdf/478073main_Day1_P03a_Pellis_Cell_Biology.pdf, (accessed 5.03.2020).

[108] Kasap, H., Exotic Glass Fibers from Space—The Race to Manufacture ZBLAN, December 2018, https://upward.issnationallab.org/the-race-to-manufacture-zblan-4-3/, (accessed 5.03.2020).

for developing the OOS industry. Satellites that have communication capabilities could continue their mission once they are refueled.

The technology of the serviced satellites in GEO would eventually become obsolete. However, their components will still be extremely valuable. The amount of cables from a satellite is extremely valuable and would represent a sustainable solution as a resource in space manufacturing. OOS is a solution to recycle launched objects.

The End of the Mission (also known as End of Life disposal, Decommissioning, or simply disposal) is an important element of the design of any space mission. Three options are possible for the disposal of satellites from Low Earth Orbit (LEO): (i) perform a controlled re-entry in the Earth's atmosphere; (ii) store the satellite in a storage orbit; (iii) reduce the satellite orbit in an orbit low enough for the atmospheric drag to be effective. The controlled reentry is the preferred LEO disposal method, the other coming with challenges for satellite traffic in commonly used orbits.[109] Disposal from GEO is different from LEO and it could be corroborated with the services offered by OOS. Decommissioned GEO satellites are retired to so-called graveyard orbits for satellites.[110] This is applicable to satellites in Geostationary orbits (GEO) which are too far away from Earth to burn in the atmosphere. Instead, the satellites closer to Earth, in Low Earth Orbit, they are redirected downward and eventually end up burning in the atmosphere.

6.14 Case Scenarios

The OOS missions could provide solutions to an identified problem, which is preventing or at least delay the functional satellites becoming space debris only because of the payload obsolete technology. In performing the OOS mission, the OOS service provider should include a set of clauses in the contract with the customer satellite operator, commercial or governmental.

The OOS services could be requested in relation to a satellite system which is registered in the country where both the OOS provider and the customer are incorporated. Another option is the satellite system to be registered by the State where the satellite operator is incorporated, which is different from the OOS service provider. Finally, another option is to request OOS services for a satellite system registered in another State than the one where the satellite operator and the service provider are incorporated. These situations need a different legal approach (see Table 6.2).

Global politics in space activities are playing an important role in developing the future framework for OOS services. The continuous militarization of space could

[109]Hull, S.M., NASA Goddard Space Flight Center—End of Mission Considerations, https://ntrs.nasa.gov/archive/nasa/casi.ntrs.nasa.gov/20130000278.pdf, (accessed 5 March 2020).

[110]International Telecommunication Union, Recommendation ITU-R S.1003, Environmental Protection of the Geostationary-Satellite Orbit, https://www.itu.int/dms_pubrec/itu-r/rec/s/R-REC-S.1003-0-199304-S!!PDF-E.pdf, (accessed 5.04.2020).

Table 6.2 Legal circumstances between the OOS customers and OOS service providers

Customer	OOS service provider
If a governmental entity is requesting the OOS service and the satellite system is registered in the same state	For the service provider to perform the OOS service, it is necessary the following A national license A contract containing – Approval from the customer to perform the service – Detailed information about the target, including registration and launching states – Protection from liability including clauses referring to an exemption of payment for the damage. A detailed clause may include the customer's agreement to give up the right to recover the costs from the OOS service provider – Intellectual property including confidentiality of data – Ownership of the satellite system during OOS performance – Clause about the dual-use free or specific approval from the government – Agreement to benefit from SSA data – Agreement to benefit from clear air space for deorbiting purposes
If a governmental entity is requesting the OOS service for a satellite system registered in another state	The OOS service provider needs the contractual approval from the registering state of the satellite system, otherwise it cannot perform the OOS service involving the satellite system registered by another State than the customer Any action could be regarded as affecting the national security of that country and could involve defense measures in accordance with article III OST and according to Chapter VII of UN CHARTER, in particular Article 51 about the right of self-defense * If approved by the registering state of the satellite system, the contractual agreement should contain the same clauses as mentioned above with the following modifications – Approval from both the customer and the other state to perform the OOS service – Protection from liability, including the agreement of both states referring to an exemption to pay the damage caused during the OOS service

(continued)

Table 6.2 (continued)

Customer	OOS service provider
If a non-governmental entity is requesting the OOS service for a satellite registered by the same state where the OOS service provider and the satellite operator are registered	The customer must provide the service provider with a clear request to service the satellite The OOS service provider must request a national license based on the request of the customer If approved by the registering state of the target, the contractual agreement should contain the same clauses as mentioned above, including – Approval from the customer – Authorization from the registration state

weaken the opposition to linking civilian space assets with defense elements.[111] In relation with the remove space debris missions it was highlighted the question "whether the doctrine of self-defence can be utilized to remove space debris in order to avoid or to reduce the risk for potential further incidents".[112]

OOS represents a high risk of collision between the OOS vehicle and the satellite system to be serviced and States may be reluctant to allow OOS as a commercial business before such an issue is clarified. This situation could raise the question if the OOS could have national security implications and if the OOS service providers could be exempted from liability. Clarification should address liability for OOS operations even before taking control of the satellite system, during approach. Addressing liability could be made through bilateral agreements in order to avoid the gaps from OST and the Convention on International Liability for Damage caused by Space Objects (LIAB).[113]

6.15 Conclusion

Data and services provided by satellite systems makes the satellites to be regarded as strategic assets. The US proved its leadership and has the capacity to lead international activities. The US is investing and developing OOS robotic missions for GEO, technology which could also contribute to space debris mitigation.

[111] Joseph Borell, at the 12th European Space Conference, https://www.euractiv.com/section/defence-and-security/news/budget-battle-hampers-eu-in-space/, (accessed 18.02.2020).

[112] Ibid. *supra* note Froehlich, A., The Right to (Anticipatory) Self-Defence in Outer Space to Reduce Space Debris, in: Froehlich, A., Space Security and Legal Aspects of Active Debris Removal, 2019, Studies in Space Policy, Springer Nature Switzerland.

[113] RES 2777 (XXVI), Convention on International Liability for Damage Caused by Space Objects, 1971, https://www.unoosa.org/oosa/en/ourwork/spacelaw/treaties/introliability-convention.html, (accessed 18.03.2020).

With the biggest ever budget on space activities and with contracting a commercial provider for the safe removal of an inactive ESA-owned object from LEO, ESA demonstrates a strong leadership for space sustainability in LEO and vision for essential new commercial services in the future.

The current envisioned business for OOS is about satellite life extension, technology updates or re-positioning of the satellites in orbit. The agreement between the satellite operators requesting the OOS service and the companies providing OOS capabilities on demand is not standardized and the contracting companies could develop their own contractual conditions, inserting clauses to protect the OOS services from liability.

The OOS activity is allowed under the space law framework and the OOS provider needs to be licensed by national authorities for each of their missions. The OOS demonstration missions are very important to prove such activity and support the development of a legal framework for commercial activities.

From an economic and business standpoint, by contracting the OOS services, the satellite revenue will cover the additional costs for satellite operators. Thus, the cost for an OOS mission should allow the satellite operator to pay for the OOS service and still have profit. However, the satellite life will still be limited and once the technology becomes obsolete, it will end up in the graveyard together with other satellites. Recycling and space manufacturing could become the object of OOS services. Extending the life, repositioning and upgrading the satellites are just a small portion from what the OOS could represent for the space community in the future. Turning Space Debris into a resource by using the materials available on non-functional artificial satellites, using satellites as spare parts for newly launched satellites or in-orbit assembly could become a profitable business. Even if LEO is gaining more attention lately, GEO satellites are a precious resource, even when non-functional.

The space law and policy role is to improve access to space to everyone, levering international cooperation. Space law and policy could prevent future disputes in outer space by contributing to the development of the necessary framework that would define, identify and restrict offensive behavior in space. If it's true that there will be no space sustainability without ADR and OOS, it is equally true that there will be no commercial activity that is not built on real and tangible international cooperation and global development of SSA and STM. The sooner will be reached a balance between commercial and military interests in outer space, the better for space sustainability and for the development of a legal framework to support the OOS.

Being a multipurpose space mission, the OOS could become a solution for ADR. Rather than focusing only on prevention, the space law framework should provide solutions in case a collision happens in outer space during an ADR or OOS mission.

The rule of law in space activities provides legal solutions to prevent situation of increasing tensions between nations. The legal framework must continue to assure the sustainable use of outer space for peaceful purposes and for the benefit of all mankind as successfully was done so far by the provisions of the Outer Space Treaty.

Cooperation in space, transparent operations and exchange of information for in orbit maneuvers are necessary for building trust and represent a successful model in the governance of space activities for OOS.

Claudiu Mihai Tăiatu is a Romanian lawyer, he graduated in 2017 from the Adv. LL.M. of Air and Space Law of the International Institute of Air and Space Law (IIASL), Leiden University, The Netherlands and in 2018 from the International Space University (ISU), Space Studies Program (SSP18). Currently, he is working with the European Space Policy Institute (ESPI) in projects related to space law and policy.

He was awarded in 2017 with the International Institute of Space Law (IISL) Prof. Dr. I. H. Ph. Diederiks-Verschoor Award for his research on Space Traffic Management. In 2018 he was awarded at the Worldwide Space Law Essay Competition "Legal Aspects Relating to Satellite Constellations" organized by the ECSL, ESPI and DLR. In 2019 he was awarded with Secure World Foundation (SWF) Scholarship to attend the International Astronautical Congress (IAC) in Washington D.C. He successfully completed several internships at the Regulatory Affairs Department of OneWeb, the International Telecommunication Union (ITU) Radiocommunication Bureau, ESPI and UNIDROIT. He is part of the SGAC Space Law and Policy Newsletter Team.

Chapter 7
On-Orbit Servicing, Opportunity and Liability

Christoffel Kotze

Abstract In terms of the Liability Convention the territory from where a spacecraft is launched will be held liable for any damages caused by any future incident on earth involving the spacecraft. In the past couple of years the general satellite "population" has been growing consistently with an even more rapid increase expected as the planned communications mega-constellations commence construction. An increase in the satellite "population" brings with it economic benefits, though it comes at a price. As the amount of objects in orbit increase so does the chance of an undesirable interaction between a functional spacecraft and one of the thousands of pieces of orbital space debris, causing failure of the spacecraft and resulting in potential damage on earth. This is especially pertinent for countries responsible for the liabilities of spacecraft, whose risk exposure profile will invariably increase. This article investigates how the use of "On-Orbit Services" could be used by parties liable under the Liability Convention to mitigate their risk load.

7.1 Introduction

The space industry is already playing a very important role in the global economy which is increasing year on year. In 2017 two large US based investment banks made predictions of the size of the space industry by 2040, on the lower end a figure of just over USD 1 trillion (Morgan Stanley) and on the higher end almost USD 3 trillion (Bank of America Merrill Lynch).[1] Figures published in 2019 version of

[1] Michael Sheetz, "The space industry will be worth nearly $3 trillion in 30 years, Bank of America predicts", CNBC 31 October 2017. https://www.cnbc.com/2017/10/31/the-space-industry-will-be-worth-nearly-3-trillion-in-30-years-bank-of-america-predicts.html (accessed February 27, 2020).

C. Kotze (✉)
NOEZ Strategic Advisory, Cape Town, South Africa
e-mail: chris@noez.co.za

© The Editor(s) (if applicable) and The Author(s), under exclusive license to Springer Nature Switzerland AG 2020

A. Froehlich (ed.), *On-Orbit Servicing: Next Generation of Space Activities*, Studies in Space Policy 26, https://doi.org/10.1007/978-3-030-51559-1_7

"The Space Report"[2] certainly does support the previously mentioned predictions; as in 2018 the space industry already had a direct economic contribution of just under USD 415 billion, with an almost 50% increase in the amount of launches as compared with those a decade before. The indirect contribution of the space industry is more difficult to determine as it has become so well-ingrained in modern society represented in virtually every socio-economic sector. The converged technology of the Fourth Industrial Revolution hungry for interconnectivity, will becoming ever more dependent on space technology; Placing increasing demands on the placement of equipment, making low earth orbit a very contested space, literally in danger of "running out of space."[3]

The effect of an increase in the satellite "population" is dualistic; on the plus side it creates conditions for a profound improvement to socio-economic conditions on earth, on the down side; proliferation satellites in low earth orbit in particular, will increase the risk of collisions between functional space objects and existing space debris.

Figure 7.1 attempts to illustrate the relative relation between amounts of tracked pieces of space debris and orbiting spacecraft both active and inactive represented by the smaller of the two "lifebuoy" graphics, the other showing low earth orbit as the preferred orbit of choice in relation to the other options. Over and above the tracked space derbis, there are literally millions of smaller pieces travelling at incredibly high speeds posing a very real danger of a sudden unexpected collision.

Space debris also colloquially known as "space junk" is generally considered to be any unwanted objects or material left in space as consequence of human activity.[4] Debris can be produced during any phase of the spacecraft's lifecycle typical; abandoned launch vehicle stages or any defunct ancillary mission related debris, failed spacecraft, end-of life spacecraft and the destruction of a craft. Destruction of spacecraft has the potential to create thousands if not millions of debris fragments travelling at high speed and is difficult if not impossible to track. Fragmentation events are normally the result of some unplanned, unforeseen or unexpected situation; however on a number of occasions these events were part of a willful action to destroy a spacecraft. A number of nation-states in order to demonstrate their strategic anti-satellite[5] capability have in the past resorted to the destruction of their own spacecraft, contributing significantly to the body of space junk. It is estimated that up to

[2]Kiona Smith, "India's Anti-Satellite Missile Test Left a Cloud of Debris and Tension in Its Wake", Forbes 5 April 2019. https://www.forbes.com/sites/kionasmith/2019/04/05/indias-anti-satellite-missile-test-left-a-cloud-of-debris-and-tension-in-its-wake/#23eb81f78fd1 (accessed March 29, 2020).

[3]Sinéad O'Sullivan, "Understanding the Space Economy", Harvard Business Review. 18 May 2019. https://www.hbr.org/podcast/2019/05/understanding-the-space-economy (accessed March 28, 2020).

[4]Jonathan O"Callaghan, "What is space junk and why is it a problem?" NHM 14 February 2019. https://www.nhm.ac.uk/discover/what-is-space-junk-and-why-is-it-a-problem.html (accessed March 19, 2020).

[5]Gerry Doyle, "Factbox: Anti-satellite weapons: rare, high-tech, and risky to test", Reuters 27 March 2019. https://www.reuters.com/article/us-india-satellite-tests-factbox/factbox-anti-satellite-weapons-rare-high-tech-and-risky-to-test-idUSKCN1R80QP (accessed March 27, 2020).

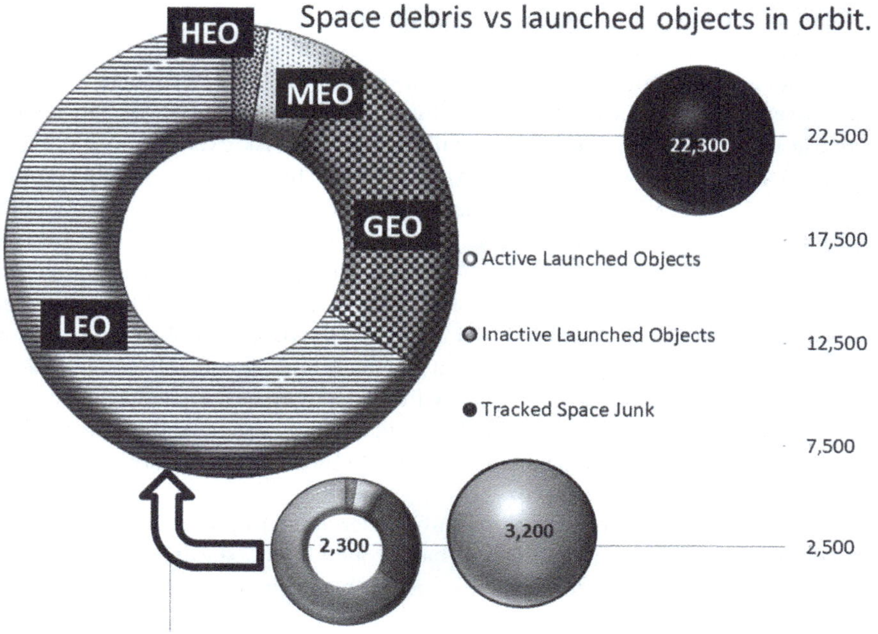

Fig. 7.1 Tracked space debris in relation to active objects in orbit

3000 pieces of space debris were formed, the largest yet, when the Peoples Republic of China destroyed one of its own weather satellites using an anti-satellite missile in 2007.[6] There are however other events as well where it might be in the interest of the liable party to destroy a satellite e.g. if it cannot be controlled anymore. In such cases, if it could be done in such a way as to ensure no pieces of the stricken satellite reach the earth, the debris risk might be the lesser of two evils. A practical example occurred in 2008 when a malfunctioning reconnaissance satellite (USA-193) was destroyed by a missile fired from the guided missile cruiser USS Lake Erie,[7] it was further claimed that a year later no debris remained as it had deorbited completely and burnt up on reentry.[8]

---

[6]Carin Zissis, "China's Anti-Satellite Test" CFR 22 February 2007. https://www.cfr.org/backgrounder/chinas-anti-satellite-test (accessed March 28, 2020).

[7]Jim Garamone, "Lake Erie Crew Describes Satellite Shot", US Navy 24 February 2008. https://www.public.navy.mil/surfor/cg70/Pages/Lake%20Erie%20Crew%20Describes%20Satellite%20Shot.aspx (accessed March 19, 2020).

[8]Jim Wolf, "U.S. satellite shootdown debris said gone from space", Reuters 27 February 2009. https://www.reuters.com/article/us-space-usa-china/u-s-satellite-shootdown-debris-said-gone-from-space-idUSTRE51Q2Q220090227 (accessed March 18, 2020).

The Convention on International Liability for Damage Caused by Space Objects[9] commonly known as the "Liability Convention" is a treaty laying out the rules for parties liable to pay compensation, in case a space object caused damage on earth. The movement of more satellites in areas inhabited by space debris will statistically increase the possibility of a collision with an elevated risk exposure to the liable parties. A strategy to lessen risk exposure thus needs to focus on how collisions could be prevented. Tactically this could be achieved in three ways; reduce the amount of satellites, prevent production of debris and reduce the active space debris. Since it is unlikely that satellite launches will be reduced, in actual fact as stated earlier the opposite is expected, the primary focus should be on debris reduction. One of the best tools available to parties interested in remediating the debris problem is the use of the developing field of "On-Orbit Servicing", which was dramatically demonstrated in early 2020. On 25 February 2020, a new frontier was opened in the space industry when a purpose designed spacecraft MEV-1 successfully extended the working-life of an active communications satellite in geosynchronous equatorial orbit which was low on fuel.[10]

This article will attempt to explore on-orbit servicing as potential liability mitigation mechanism, against a number of background topics, starting with, what the liability convention entails, risk and the current orbital situation with a brief analysis of current on-orbit servicing initiatives.

7.2 Liability Convention

The Liability Convention provides procedures on how claims are to be settled in case of damage to "health and wealth" on earth (inclusive of aircraft) by falling debris caused by failed space objects. In short the country which owns the territory from which the space objects are launched will be deemed internationally liable for any future damage on earth caused by the space objects in question. The liability can also be extended to include more than one nation state in cases where different countries collaborated in the launch of a space object. Compensation claims lodged under the Liability Convention can only happen at a state level i.e. the aggrieved parties will have to be represented by their respective countries in making claims against the state/s responsible liable for the damage. The Liability Convention acts in conjunction with any existing legislation at national level regarding compensation in relation to the damage caused.

[9]UN, "Convention on International Liability for Damage Caused by Space Objects", Vienna United Nations, 1971.

[10]Elizabeth Howell, "Two private satellites just docked in space in historic first for orbital servicing", Space 27 February 2020. https://www.space.com/private-satellites-docking-success-northrop-grumman-mev-1.html accessed March 22, 2020).

One of the most well-known examples of application of the Liability Convention is that of "Kosmos 954"[11] an intelligence gathering satellite launched by the now defunct Union of Soviet Socialist Republics in 1977. This nuclear powered satellite malfunctioned not long after it was launched and reentered the earth's atmosphere late January of 1978. The craft was equipped with a special safety mechanism to prevent the reentry of the reactor core by ejecting it into a so-called safe "graveyard orbit", this process failed though and as the stricken craft broke up in the atmosphere, it covered a vast area of Northern Canada with debris including parts of the radioactive reactor core. A cleanup operation called "Operation Morning Light"[12] was launched but only recovered a very small amount of the radioactive debris. Ultimately the USSR paid a compensation of three million Canadian dollars in terms of a settlement reach in terms of the Liability Convention in 1981.[13]

As far as collisions in space between objects, such as a piece of debris and a functional satellite are concerned, the question of determination of the liable party becomes more opaque as there is no legal definition of space debris as yet, though there is a formalized definition.[14] In 2007 United Nations Committee on the Peaceful Uses of Outer Space (UNCOPOUS) adopted a set of guidelines put forward by its Science and Technology Subcommittee to formalize the definition of space debris as: "All man-made objects including fragments and elements thereof, in Earth orbit or re-entering the atmosphere, that are non-functional."[15]

7.3 Risk—A Brief Background

Though there are many interpretations of what risk entails, generally it revolves around the chance of an adverse event happening and impact thereof. The Canadian Centre for Occupational Health and Safety defines risk as; "The combination of the likelihood of the occurrence of a harm and the severity of that harm."[16] For any individual or organization exposed to risk it is important to understand their risk exposure and ways to deal with it generally known as risk management.

---

[11] Alexander Cohen, "Cosmos 954 and the international law of satellite accidents", Yale J. Int'l L. 10 1984:78.

[12] Steve Weintz, "Operation Morning Light: The Nuclear Satellite that Almost Decimated America", National Interest 23 November 2015. https://www.nationalinterest.org/feature/operation-morning-light-the-nuclear-satellite-almost-14411 (accessed March 29, 2020).

[13] "Settlement of Claim between Canada and the Union of Soviet Socialist Republics for Damage Caused by Cosmos 954" JAXA. 2 April 1998. https://www.jaxa.jp/library/space_law/chapter_3/3-2-2-1_e.html (accessed March 30, 2020).

[14] Scott Kerr, "Liability for space debris collisions and the Kessler Syndrome", Space Review December 2017. https://www.thespacereview.com/article/3387/1 (accessed March 31, 2020).

[15] "Space Debris Mitigation Guidelines of the Committee on the Peaceful Uses of Outer Space", as annexed to UN doc. A/62/20, Report of the COPUOS 2007.

[16] "Hazard and Risk". https://www.ccohs.ca/oshanswers/hsprograms/hazard_risk.html (accessed March 25, 2020).

Table 7.1 General categories of risk

	Risk categories
Preventable	Typically internal to the organization, which can either be controlled or simply eliminated from the outset
Strategic	Risks that an organization accepts in order to achieve an objective, these have to be managed carefully
External	These exist outside the organization's sphere of influence and cannot be controlled therefore appropriate mitigation measures need to be identified

Risk management is one of the most researched aspects of the modern economic landscape, a very complex process with no easy answers where quite often, "The road to hell is paved with good intentions."[17] The aforementioned English proverb conveys the message that it is not enough to just mean well without taking the appropriate action to actually do well, is in practice quite applicable to many failed risk management processes. Sadly there are many examples of risk management failure which can be found in the business world, some of such a magnitude as to cause systemic damage. In 2007 the so-called "sub-prime" banking crises set in motion the 2008 financial crises leading to the worst recession since the "Great Depression" at the time.[18] Other examples of risk failures not only resulted in financial losses but also in lasting environmental damage. In 2010, "Deep Water Horizon", an offshore oilrig exploded causing the spillage of more than 800 million liters of oil causing significant environmental damage with the knock-on effects still felt today.[19]

In a 2012 Harvard Business Review article Kaplan & Mikes[20] cited the "Deepwater Horizon" tragedy as an example of a risk management failure caused by organizations focusing on a compliance and rules-based risk management approach which resulted in the failure to identify crucial risks unique to the operational situation.

They classified three broad categories of risk that organizations should interrogate against their unique realities, be it at enterprise or project level, namely (Table 7.1):

To manage risks an organization will determine its "Inherent Risk" in other words the basic level of "raw" risk the organization is exposed to. The nature and level of the inherent risk will normally be the main influence in defining a strategy to manage the risk. For the purpose of this document the definition of risk management used by the Association for Project Management will be used i.e. "A process that allows

[17] Izzy Kalman, "Principle One: Road to Hell is Paved with Good Intentions", Psychology Today 16 August 2010. https://www.psychologytoday.com/za/blog/resilience-bullying/201008/principle-one-road-hell-is-paved-good-intentions (accessed March 28, 2020).

[18] Kimberly Amadeo, "Subprime Mortgage Crisis, Its Timeline and Effect." The Balance. 20 November 2019. https://www.thebalance.com/subprime-mortgage-crisis-effect-and-timeline-3305745 (accessed March 17, 2020).

[19] Vaughan Adam, "Deepwater Horizon spill may have been a third bigger than estimated", NewScientist, 12 February 2020. https://www.newscientist.com/article/2233346-deepwater-horizon-spill-may-have-been-a-third-bigger-than-estimated/ (accessed March 17, 2020).

[20] Robert Kaplan and Anette Mikes, "Managing risks: a new framework", 2012, Harvard business review 90, no. 6, pp 48–60.

Table 7.2 Risk management strategies

	Risk management strategies
1	Avoidance—the choice is made to simply not engage in any activity that may lead to a specific risk and the subsequent liability
2	Acceptance—the potential risk is acknowledged and any associated liability is identified
3	Reduction—the liable party puts in place measures to reduce the severity of the risk
4	Transfer—the liable party passes on the risk burden either in whole or partly to another party

individual risk events and overall risk to be understood and managed proactively, optimizing success by minimizing threats and maximizing opportunities."[21] The choice of risk management strategy will be further determined by two other influencers namely the organization's "risk-appetite" and "risk-tolerance". The risk-appetite is seen as a high level boundary determined by an organization to indicate the amount of risk it is willing to take on, variances around which is known as risk tolerance.[22] In order to reduce its residual risk in line with its determined risk appetite an organization will define a general risk management approach. Risk management typically will involve one of four basic broad approaches as set out in Table 7.2.[23]

After the organization has implemented its chosen risk mitigation plan normally involving combinations of the general risk management strategies, it will typically still be left with a certain amount of risk generally known as the "Residual Risk".

Parties exposed to potential litigation for damages caused by spacecraft as per the Liability Convention, will typically be exposed to a risk profile consisting of a combination of those mentioned in the previous section Table 7.1. A large portion of the risk component will be skewed towards the external category leaving these parties with high levels of residual risk. Since the mitigation objective of the parties liable for spacecraft is fairly straightforward; minimize the chance of a satellite entering the earth's atmosphere in such a way that large pieces still not survive the reentry and thus exposing them to claims. As is not possible to control external risks in space and the risk cannot be transferred to any other party, the risk management strategy therefore needs to be adaptable and creative using a combined approach of strategies. Importantly it will strive to identify the known dangers and then explore risk mitigation strategies skewed towards reduction. Take the following two scenarios involving a spacecraft in low Earth orbit and the proximity of space debris:

[21]"Introduction to Risk management". https://www.apm.org.uk/body-of-knowledge/delivery/risk-management/ (accessed March 27, 2020).

[22]"Risk Appetite versus Risk Tolerance, What's the Difference?" FAIR Institute 1 May 2017. https://www.fairinstitute.org/blog/risk-appetite-vs.-risk-tolerance.-whats-the-difference (accessed March 10, 2020).

[23]Melissa Horton, "Common Examples of Risk Management", Investopedia. 20 December 2018. https://www.investopedia.com/ask/answers/050715/what-are-some-examples-risk-management-techniques.asp (accessed March 14, 2020).

Scenario 1: Spacecraft is threatened by a piece of debris that is tracked and ground control has identified that the two objects are in a possible collision trajectory. In this scenario, the risk can be mitigated by:

- Acceptance—leave the craft in position and accept the consequence and resulting liability if a collision occurs.
- Avoidance—temporarily repositioning the spacecraft to a safe environment to prevent a possible collision, though there will be a tradeoff as the action will burn fuel and thus ultimately lessen the operational lifetime of the craft.
- Reduction—move the satellite to a new orbit free of tracked debris where it can still function.
- Transfer—insure against possible damages through an insurance company and claim in case of a collision.

Scenario 2: Spacecraft is threatened by a piece of debris that is not tracked and no one is aware of the fact that the two objects are in a possible collision trajectory. In this scenario since the specific threat condition is not identified mitigation will be based on that for residual risk, leaving very little options really:

- Acceptance—in the case of a collision event accept the consequence and the resulting liability.
- Transfer—insure against possible damages through an insurance company and claim in case of a collision.

7.4 Orbital Traffic and Hazards

Human exploration of space was initiated on 4 October 1957 with the launch of a spherical object roughly the size of a modern Pilates exercise ball, named Sputnik, since then thousands of objects have launched and accordingly registered with the United Nations Office for Outer Space Affairs.[24] With close to 2500 active satellites expected to be in orbit by mid-2020, and approval for the launch of thousands of additional satellites already granted by the Federal Communications Commission in the USA, space is becoming contested especially in low earth orbit increasing the collision hazard significantly.[25] The infographic depicted in Fig. 7.2 based on data compiled by the ESA's European Space Operations Centre published in February 2020, attempts to communicate the breakdown of the total mass in orbit around earth, estimated to be in excess of 8800 tons.

According to current estimates more than half a million individual pieces of space junk orbits the earth travelling at incredible speeds that can exceed 28,000 km per

[24]UN on-orbit servicing, Online Index of Objects Launched into Outer Space. https://www.unoosa.org/oosa/osoindex/search-ng.jspx?lf_id= (accessed March 27, 2020).

[25]Caleb Henry, "SpaceX submits paperwork for 30,000 more Starlink satellites", SpaceNews 15 October 2019. https://www.spacenews.com/spacex-submits-paperwork-for-30000-more-starlink-satellites/ (accessed March 27, 2020).

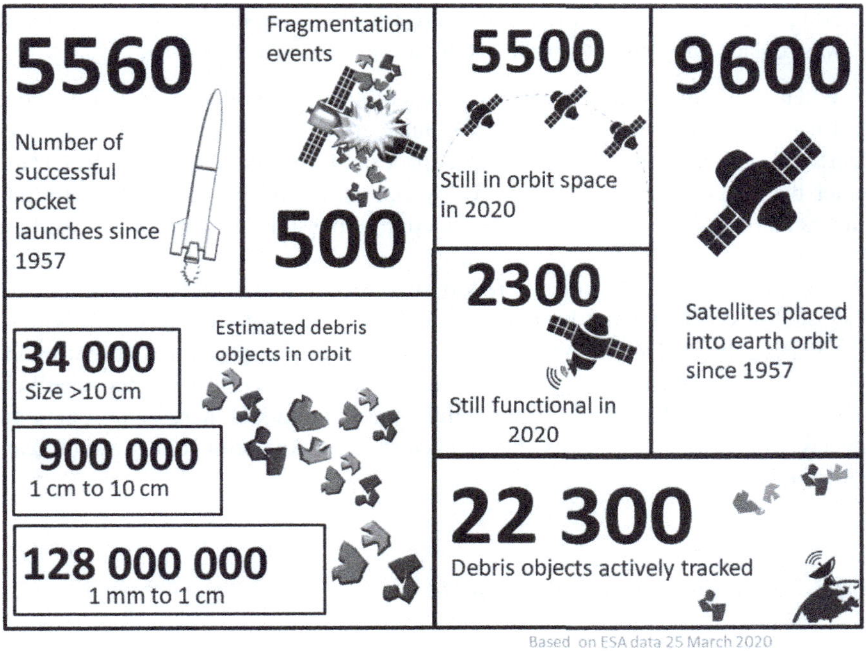

Fig. 7.2 Space debris infographic based on 2020 ESA data. ["Space debris by the numbers ESA February 2020". https://www.esa.int/Safety_Security/Space_Debris/Space_debris_by_the_numbers (accessed March 17, 2020)]

hour of this group a collection of about twenty thousand is larger than a baseball.[26] A debris item in the aforementioned group colliding with a function spacecraft can easily destroy the craft or at best in the case of a large craft such as the ISS cause considerable damage. A number of events have proved this point;

- 1996 the surveillance satellite "Cerise"[27] belonging to the French collided with a piece of catalogued debris, a leftover from the explosion of an Ariane 1 launch vehicle's third stage in 1986.[28]
- 2009 a fully functional Iridium communications satellite was destroyed by an old Russian satellite in the process both objects were destroyed producing more than

[26]Mark Garcia, "Space Debris and Human Spacecraft", NASA 27 September 2013. https://www.nasa.gov/mission_pages/station/news/orbital_debris.html (accessed March 29, 2020).

[27]Mark Ward, "Satellite injured in space wreck" New Scientist 24 August 1996. https://www.newscientist.com/article/mg15120440-400-satellite-injured-in-space-wreck/ (accessed March 31, 2020).

[28]Spaceref. Accidental Collisions of Cataloged Satellites Identified. 16 April 2005. https://www.spaceref.com/news/viewsr.html?pid=16201 (accessed April 15, 2020).

500 pieces of debris.[29] A point of interest in this case, the lost Iridium satellite was replaced by one of the spares kept in orbit for such an occasion.[30]

When a satellite is destroyed by either an unplanned collision or and intended destruction by the owner it has the unfortunate effect of producing additional space debris as previously discussed. The more space debris present the greater the hazard, either by direct collision or indirectly by deflecting another piece of debris into causing, albeit unintended, the loss of function or possible destruction of a working spacecraft.

According to a 2016 forecast report it is expected by 2025, an additional 9 000 satellites would have been placed in orbit, the majority of which will be low Earth orbit.[31] This number will include the initial operational stages of the proposed communication mega-constellations including Project Kuiper, OneWeb, O3b mPOWER and StarLink. The combination of a marked increase in functional objects and thousands of pieces of untracked space junk in orbit creates a significant collision hazard.

7.4.1 In-Orbit Collision Hazards

The larger pieces of space junk which are tracked does allow in most cases for the owners of a spacecraft to effect evasive maneuvers to avoid collisions. Take the example of the CryoSat-2 used to monitor ice cover on earth and is operated by the European Space Agency in low Earth orbit at 700 km. This satellite was successfully repositioned on 2 July 2018, less than an hour before it was about to collide with a piece of debris.[32] It must also be remembered that there are many (in the millions) pieces of space debris that cannot be tracked simply due to its small size. A very small particle travelling at high speeds can cause a lot of damage to a large spacecraft. In 2016 a 7 mm-diameter chip in a cupola window of the International Space Station was caused by nothing more than a paint fleck believed to be travelling at just under

[29] Andrew Moseman, "U.S., Russian Satellites Crash 400 Miles over Siberia", Popular Mechanics 18 December 2009. https://www.popularmechanics.com/space/satellites/a4190/4303472/ (accessed March 29, 2020).

[30] Ian O'Neill, "Orbital Spares: Iridium Already Replaced Destroyed Satellite", Universe Today 14 February 2009. https://www.universetoday.com/25447/orbital-spares-iridium-already-replaced-destroyed-satellite/ (accessed March 29, 2020).

[31] "Satellites to be Built & Launched by 2025 Report—Analysis, Technologies & Forecasts—Market Worth USD 250 Billion by 2025", Business Wire 20 September 2016. https://www.businesswire.com/news/home/20160920005880/en/Satellites-Built-Launched-2025-Report---Analysis (accessed March 30, 2020).

[32] Alexandra Witze, "The quest to conquer Earth's space junk problem", Nature 5 September 2018. https://www.nature.com/articles/d41586-018-06170-1 (accessed March 16, 2020).

35,000 km per hour.[33] Such collisions between orbiting objects usually take the form of a hypervelocity impact (HVI), the effect of which could be far reaching. According to the ESA a HVI[34] is defined as a collision between space objects where the relative velocity exceeds 4000 m/s and can even be much higher, in the region of 20,000 m/s for a spacecraft encountering a meteoroid. At those speeds the effect caused by HVI will primarily be determined by the size of the "impactor" (the object creating the damage) and where the impact takes place on the target. Subsequently damage may take many forms for example:

- A crewed spacecraft is punctured endangering the inhabitants.
- Spacecraft is destroyed by an explosive collision leading to a lot of space debris being formed.
- Spacecraft is functionally impaired but the orbit remains intact.
- Spacecraft becomes uncontrollable due to a loss of fuel or other control mechanisms making the damaged craft a potential impactor itself.
- Spacecraft's orbit is adversely influenced which if not corrected can lead to an uncontrolled atmospheric reentry causing damage on earth.

It is estimated that 50% of the annual collision avoidance maneuvers performed by the ESA fleet of satellites can be attributed to hazards created by the fragmentation of spacecraft such as the destruction of its own craft by China as previously mentioned.[35] Space debris pose a definite hazard for any orbiting space craft especially in low Earth orbit and could potentially present a much more devastating situation if a scenario postulated by Donald Kessler in 1978 were to occur.

7.4.2 Kessler Effect

Named after its originator, the "Kessler Effect" refers to situation where the presence of space junk can potentially create a cataclysmic chain-reaction to turn working satellites in low Earth orbit into a swarm of orbiting debris in 1978.[36] Kessler a

---

[33] Ellie Zolfagharifard, "What happens when a tiny fleck of paint hits the space station: Tim Peake reveals crack in ISS window after debris collides with craft". https://www.dailymail.co.uk/sciencetech/article-3587882/What-happens-tiny-fleck-paint-hits-space-station-Tim-Peake-reveals-crack-ISS-window-debris-collides-craft.html (accessed March 16, 2020).

[34] "What are hypervelocity impacts?". https://www.esa.int/Enabling_Support/Operations/What_are_hypervelocity_impacts (accessed March 17, 2020).

[35] Kiona Smith, "India's Anti-Satellite Missile Test Left a Cloud of Debris and Tension in Its Wake", Forbes 5 April 2019. https://www.forbes.com/sites/kionasmith/2019/04/05/indias-anti-satellite-missile-test-left-a-cloud-of-debris-and-tension-in-its-wake/#23eb81f78fd1 (accessed March 29, 2020).

[36] Michelle LaVone, "The Kessler Syndrome: 10 Interesting and Disturbing Facts", Space Safety Magazine 17 May 2009. https://www.spacesafetymagazine.com/space-debris/kessler-syndrome/ (accessed March 29, 2020).

NASA scientist at the time postulated a future scenario where low Earth orbit could be totally filled with an orbiting "debris-cloud". How can such a scenario unfold; The process will be initiated with the collision of two objects in orbit, which in turn will create some debris, which in turn creates more opportunity for collision and thus more debris culminating in a perpetual "debris collision cascade". The process will eventually produce a debris cloud so dense that at some stage it could prevent the positioning of new working satellites in higher up orbits. Such a scenario needs to be prevented at all cost as the space industry has become an indispensable part of the modern socio-economic system. It plays a significant role in the USD 5 trillion per annum agriculture and food industry by enabling precision-agritech ensuring profitability and sustainability of a sector which will be challenged increasingly by critical resource shortages most notably water.[37]

7.5 On-Orbit Services

On-orbiting services (on-orbit servicing) describe activities that can be applied in orbit to create an environment more conducive to supporting sustainable use of space. These activities can be wide ranging including but not limited to refuelling, repairing, preventative maintenance, technology upgrades and importantly making the operating environment safer by cleaning up debris.

The Space Shuttle[38] program which was operated by NASA from 1981 to 2011 was primarily designed as a reusable multifunctional vehicle to carry cargo, launch space objects and also serving as an orbiting laboratory. Lesser known perhaps it was also applied in an on-orbit servicing role on a number of occasions. Most well-known perhaps was the application of a solution mitigating a design flaw of the Hubble Space telescope (discovered on deployment), the first-ever on-orbit repair.[39] The shuttle subsequently revisited Hubble on a number of occasions in service missions for routine repair and maintenance and technology upgrades. With a total of fourteen fatalities recorded during the Shuttle program with the loss of the Challenger and Columbia, the shuttle was also a stark reminder of the perils associated with manned missions.[40] Advances in robotic and ancillary supporting technologies such as artificial intelligence (machine learning and deep learning) and sensors are providing new options to create effective autonomous on-orbit services which can rather support manned missions as opposed to rely on it for on-orbit servicing. The following section

[37] "How Space Data is Enabling the Agritech Sector". https://www.business.esa.int/news/how-space-data-enabling-agritech-sector (accessed March 25, 2020).

[38] Sarah Loff, "The Space Shuttle Era", NASA 4 August 2017. https://www.nasa.gov/mission_pages/shuttle/flyout/index.html (accessed March 29, 2020).

[39] Rob Garner, "About—Hubble Servicing Missions", NASA 7 February 2020. https://www.nasa.gov/mission_pages/hubble/servicing/index.html (accessed March 31, 2020).

[40] Carol Pinchefsky, "5 Horrifying Facts You Didn't Know About the Space Shuttle", Forbes 18 April 2012. https://www.forbes.com/sites/carolpinchefsky/2012/04/18/5-horrifying-facts-you-didnt-know-about-the-space-shuttle/#35cf3652f9d4 (accessed March 14, 2020).

Table 7.3 Cost of placing a kg of spacecraft in place [Wendy Whitman Cobb, "How SpaceX lowered costs and reduced barriers to space", The Conversation 1 March 2019. https://www.theconversation.com/how-spacex-lowered-costs-and-reduced-barriers-to-space-112586 (accessed March 29, 2020)]

	Cost to put a kg of material into orbit
Period from 1970 to 2000 average	USD 18,500 per kg
Space shuttle	USD 54,500 per kg
Falcon 9 trip to ISS	USD 2720 per kg

will briefly explore how on-orbit servicing may provide support to a sustainable space economy using a popular framework borrowed from the "Green Economy" colloquially known as the "Four R's"[41]; rethink, reduce, reuse and recycle.

7.5.1 Rethink

Rethinking the status quo is the logical point of departure when considering sustainability options in any sector. How can current unsustainable processes be done differently to not only solve the problem it currently creates but also to add value by creating a future sustainable environment. Rethink is arguably the most powerful "R" as it more often than not creates a cripple effect of support for the other three "R's." So how can it be applied to on-orbit servicing?

Perhaps one of the most well-known examples of rethinking a "mature" process was the introduction of reusable launch systems by SpaceX. Involving extremely high costs to place a satellite in orbit, makes launch systems a good place to start. Launch cost has been extremely high for most of the duration of the space economy that is until SpaceX introduced a more sustainable approach to launching through the introduction of reusable launch vehicles. Though technically the Space Shuttle was the first attempt at a reusable launch vehicle the costs were prohibitively high. Putting it into perspective to put one kg in orbit using the Shuttle was roughly 2.5 times more expensive than the standard launch systems at the time and compared to today, almost twenty times more expensive than the SpaceX Falcon 9 system refer Table 7.3. The success of the SpaceX reusable program where care is taken to reuse as much of the launch system as possible, has inspired other companies to follow suite including Blue Origin, Orbital ATK and the United Launch Alliance which may ultimately lead to make launching much more affordable and sustainable.

Another major challenge in the role-out of space technology thus far has been "the mission approach" involving a bespoke product with a well-defined purpose and finite lifetime. These were never designed with the intention to be repaired or

[41]Tim Laseter, Anton Ovchinnikov, and Gal Raz, "Reduce, Reuse, Recycle...or Rethink", Strategy Business 23 November 2010. https://www.strategy-business.com/article/10406?gko=fd9f7 (accessed March 28, 2020).

serviced in orbit, essentially a throwaway philosophy. In rethinking this approach the satellite design architecture needs to be changed and replaced with one that will support future on-orbit servicing. An ideal architecture should not only ensure the future maintenance, repair and upgrade of the spacecraft in orbit but also for the complete construction of a new spacecraft in space. By creating functional objects in space not only can launch costs from earth can be limited to component supply runs, but any limitations imposed on the design by the constraints of the launch process can be eliminated. The "Phoenix"[42] program of the USA's Defense Advanced Research Projects Agency explored such a concept though the use of low cost modular building blocks called "Satlets" that could be used to assemble a variety of functional satellites in low earth orbit.

7.5.2 Reuse

Reuse, as the term implies is an attempt is made to gain the maximum use of a product by using it more than once to a point where it needs to be discarded e.g. liquefied petroleum gas cylinders can be refilled many times provided it still conforms to the relevant safety standards. The reusable launch systems mentioned earlier are obvious examples of reuse and its positive impact, in March of 2020 SpaceX used a Falcon 9 rocket for a record fifth launch practically demonstrating what can be achieved.[43] By making it cheaper to launch objects to perform on-orbit services, it can assist in sustainability as the following examples try to illustrate:

- Satellites that have come to the end of their functional life typically in geosynchronous equatorial orbit can have its lifetime extended by being refuelled; the International Space Station has been a regular recipient of refuelling missions to ensure its continuous operation.
- Another subset of reuse would be repurposing i.e. after its initial useful life has ended for example a retired shipping container is repurposed to create a residence for someone. In a future where a high latency communications satellite might no longer be relevant, it could be upgraded to serve a different purpose, similarly so a retired communications satellite in geosynchronous equatorial orbit might be repurposed to assist in space debris clean-up operations. The on-orbit servicing missions conducted by NASA to restore the Hubble space telescope to full function after its initial partial failure and later technology upgrades can needless to say also be seen as falling into the repurpose category.

[42] Todd Master, "DARPA Phoenix". https://www.darpa.mil/program/phoenix (accessed March 31, 2020).

[43] Loren Grush, "SpaceX successfully launches the same rocket for the fifth time but doesn't stick the landing", The Verge 18 March 2020. https://www.theverge.com/2020/3/17/21183334/spacex-falcon-9-rocket-starlink-launch-5th-time (accessed April 1, 2020).

7.5.3 Reduce

Lessening the impact of a negative agent such as waste, perhaps the most well-known example on earth is the attempts at reducing greenhouse-gasses in an effort to reign in climate change.[44] In space the reduction of space debris is a high priority to ensure the sustainably of space for economic use by preventing the Kessler effect, this can be achieved through on-orbit servicing as illustrated by these examples:

- In 2019 ESA announced it commissioned ClearSpace[45] to conduct a space debris removal demonstration mission (ClearSpace-1) involving the removal of the ESA's own Vega Secondary Payload Adapter from low earth orbit in 2025.
- A similar initiative was announced by Japan Aerospace Exploration Agency appointing Astroscale a debris removal company to remove a Japanese launch vehicles' spent upper stage in orbit.[46]

7.5.4 Recycle

Taking the waste products of one process and converting it into something useful for another process e.g. turning food waste into compost for use as fertilizer. This cycle normally involve the collection of junk, placing it in a central repository e.g. refuse centre from where it can be sorted; any valuable material recovered for recycling and separated from those of no value which in turn should be destroyed through incineration. A similar path is suggested using on-orbit services as seen in the following example:

- A proposal by Northrop Grumman is analogous to the terrestrial junk recycling route to deal with the space debris problem. The system involves the placement of an on-orbit recycling system which can be used to not only track down space junk, but also recover any useful material. Waste material will be recycled where possible or simply got rid of through incineration using the energy of the sun in conjunction with a concentrating mirror.[47] Recycled materials could be melted and stored as input for space based manufacturing, it is suggested that the system

---

[44] "Climate Change Indicators: Greenhouse Gases", EPA 22 February 2017. https://www.epa.gov/climate-indicators/greenhouse-gases (accessed March 31, 2020).

[45] Szondy, David, "World-first space debris removal mission to launch in 2025", New Atlas 9 December 2019. https://www.newatlas.com/space/world-first-space-debris-removal-mission-to-launch-in-2025/ (accessed March 19, 2020).

[46] Darrell Etherington, "Orbital debris start-up Astroscale chosen by JAXA for its first space junk removal mission". https://www.techcrunch.com/2020/02/12/orbital-debris-startup-astroscale-chosen-by-jaxa-for-its-first-space-junk-removal-mission/ (accessed April 1, 2020).

[47] Brooks McKinney, "How to Recycle Space Junk and Reduce Launch Costs" 25 March 2019. https://now.northropgrumman.com/recycle-space-junk-reduce-launch-costs (accessed April 1, 2020).

be mad self-sustaining to a certain extent by using recycled material for aspects of its own maintenance e.g. resurfacing its mirror by melting aluminium debris.

- Most redundant geosynchronous equatorial orbit satellites currently retire to the "graveyard"[48]—an orbit situated a couple of hundred kilometres above geosynchronous equatorial orbit—these satellites can potentially be scavenged for useful parts such as PV arrays and structural components which could in turn be used to construct new structures in space e.g. to serve as waystations or assembly platforms.

7.5.5 On-Orbit Servicing—The Basics

Though on-orbit service missions might differ in their intended purpose i.e. debris removal versus refuelling, the steps involved in reaching the respective end-goals will be fairly generic. It will involve a service craft of an advanced nature featuring applicable state-of-the-art technology including; artificial intelligence, intelligent pervasive sensors and robotics arm systems with specialised interchangeable end-of-arm toolsets. Such a craft must be able to successfully approach and engage its target in close proximity, dock with it to complete its mission and subsequently return the target spacecraft back to service all without compromising the integrity of target and service craft itself. A high degree of autonomous capability will be required during the mission as remote control from earth will not be possible due to latency issues for any time dependent activity such as grappling the target craft.

A brief analysis of the planned Restore-L[49] on-orbit service project can be used to provide more detail of an on-orbit service mission. Restore-L is a NASA project featuring a robotic spacecraft that will have the ability to refuel and service existing spacecraft. Scheduled to launch late in 2022, it is intended to rendezvous with the USD 500 million low Earth orbit observation satellite Landsat 7 to demonstrate its abilities.[50] During this visit the service satellite will make alterations to Landsat 7 to ensure easy future refuelling sessions through the fitment of the necessary equipment as well as actually refuelling the craft. One of the most challenging aspects of the intended procedure is that Landsat 7 was never designed to be refuelled, as is the case with most of the satellites in orbit at the current moment. Restore-L is considered to be very significant for future on-orbit servicing development as the design of the target satellite is considered to be very representative nature. The mission will involve a combination of procedures using remote operation from earth and fully

---

[48]"Graveyard Orbits and the Satellite Afterlife", NOAA 31 October 2016. https://www.nesdis.noaa.gov/content/graveyard-orbits-and-satellite-afterlife (accessed March 31, 2020).

[49]Jeffrey Hill, "Maxar to Ship NASA"s Restore-L Spacecraft in 2020", Satellite TODAY 8 April 2019, https://www.satellitetoday.com/innovation/2019/04/08/maxar-to-ship-nasas-restore-l-spacecraft-in-2020/ (accessed March 31, 2020).

[50]Evan Ackerman, "How NASA Will Grapple and Refuel a Satellite in Low Earth Orbit" IEEE 2 October 2019. https://www.spectrum.ieee.org/tech-talk/aerospace/satellites/how-nasa-will-grapple-and-refuel-a-satellite-in-low-earth-orbit (accessed March 12, 2020).

Table 7.4 The proposed 2025 On-orbit service to refuel Landsat 7

Step	Description
1	Engage the target by grappling the "Marman" ring it with a robotic arm; A structure used to connect the satellite to the upper-stage, a standard feature on most satellites
2	Using a dedicated tool an incision is created through the protective layer to access the fill and drain valves
3	To gain access to the valves the locking wires need to be severed a complicated procedure which will be conducted using the robotic arm and another dedicated tool
4	Valve caps are unscrewed
5	Equip the fuel valve with a "quick disconnect" port which will make future refuelling a less complicated procedure
6	Refuel the satellite—in the case of Landsat involving the transfer of 15 kg hydrazine fuel
7	Patch up all areas where protective material was removed with thermal blankets

autonomous actions. Latency issues leading to delayed responses can make certain procedures not suitable for remote control from Earth, necessitating full autonomous control e.g. the grappling procedure. Arguably the most important action will be the initial approach and grappling of the craft, as both spacecraft will be traveling at very high speeds, delayed reaction due to time lags could easily result in a mistake, resulting in the loss of both spacecraft, thus it will need to be a fully autonomous operation. Table 7.4 describes the seven key steps involved in the envisioned process, before the target is returned to service, an extended service that is.

The process described above can very easily be adjusted for another purpose through the use of different tools but in principle following the same steps. It is conceivable that as the process evolves in practice the emergence of a multifunctional craft will become the norm.

7.6 Conclusion

This work attempted to look at on-orbit services in its role as potential risk mitigation agent for parties liable under the Liability Convention. This was done against the background of a rapid proliferation of satellites in low earth orbit and the increasingly high levels of space debris, the proximity of both creating an increased collision risk. On-orbit servicing is a novel solution that can be applied to this problem in a two-pronged strategic approach; remove debris on the one hand and on the other extend the lifetime of existing satellites. On-orbit solutions can add additional value over and above that derived from solving the initial problem by recycling debris into raw material for future on-orbit assembly. On-orbit servicing thus holds incredible potential, however, having said so there is potential cause for concern.

Why so, well sometimes introducing a novel solution to a unique problem can in itself create an even bigger though unintended problem, examples of which are unfortunately easy to find. Australia's cane toad problem is a case in point, originally introduced as a novel solution to protect sugarcane from insect pests it soon became a much bigger and more widespread problem in itself.[51] Like the cane toad analogy, care must be taken that the introduction of on-orbit services does not create a greater debris problem, the very problem it was designed to eliminate. On-orbit services will involve fairly risky close-proximity procedures between two orbiting objects taking place at very high speed, any failure can easily result in a collision leading to a fragmentation event. Preventing it could be a very difficult task; one suggestion could be a primary rule, something in the vein of Asimov's laws of robotics. In his 1942 short story "Runaround"[52] Isaac Asimov described a number of hypothetical 'laws' which would govern the interaction of robots with their human masters. Maybe a similar set of rules should be included as foundation algorithm to govern the relationship between any robotic on-orbit service craft and its service environment independent of its mission i.e. *never through wilful action or through inaction cause space debris.* Though the aforementioned is a very basic attempt at a simplified solution to a potentially complex problem, it is something that needs consideration. Though the time where smart autonomous on-orbit service craft will crisscross earth orbit in search of debris and maintenance runs might be still a while away their potential benefit could be outweighed by their risk if not managed proactively.

In conclusion it needs to be accepted that the modern world is much more reliant on space technology than is generally acknowledged, without the enabling support of space technology the momentum of transformative human development movements such as the Fourth Industrial Revolution can be significantly curtailed. As such a blow to the space economy is a blow to the global socio-economic system and space debris has potential to plant a knock-out blow in that it can start a collision cascade known as the Kessler effect. Such a scenario will at best make it very expensive to launch satellites and at worse, make low earth orbit an unusable "no-go" zone, causing a systemic socio-economic implosion on earth. Such a situation should be avoided at all cost.

Christoffel (Chris) Kotze established a boutique technology strategic advisory company in 2012 after a successful corporate career spanning two decades. This company specialises in providing assistance to Digital Transformation projects within organizations, with a special interest in the use of technology resources to support sustainable development. Current research interests include space technology, dematerialisation through digital transformation and solutions to the "digital divide". Qualifications include M.Phil. (Space Science) University of Cape Town, Bachelor of Commerce Honours (Information Systems)—University of Cape Town, Bachelor of Science (Physiology & Microbiology)—University of Pretoria, Diploma in DataMetrics (Computer Science) University of South Africa, a number of strategy focussed executive management courses at the Graduate School of Business from the University of Cape Town. ISACA Certified in the Governance of Enterprise IT (CGEIT), TOGAF 9 Certified (Enterprise Architecture).

---

[51] Tina Butler, "Cane toads increasingly a problem in Australia", 17 April 2005. https://news.mongabay.com/2005/04/cane-toads-increasingly-a-problem-in-australia (accessed April 5, 2020).

[52] Isaac Asimov, "Runaround" Astounding Science Fiction (1942), 19(291), pp 94–103.

Chapter 8
On-Orbit Servicing from a Legal and Policy Perspective

Margaux Morssink

Abstract As the economic success of private actors in outer space keeps rising, sustainable use of outer space is becoming a more pressing issue. Although mitigation of space debris and sustainability of outer space activities have been issues for a long time now, there has never been a real impetus for the private space actors to get involved, as financial benefits often outweigh environmental considerations. As more and more satellites are launched, on-orbit servicing missions could provide for a more sustainable way of conducting space activities. Past missions have shown that it is technologically possible to conduct on-orbit servicing missions. The author argues that the solution to making on-orbit servicing a success is entering into public–private partnerships. States should initiate these partnerships and thus create an incentive for private actors to take on responsibility and accountability for the activities they undertake in outer space.

8.1 Introduction

The past couple of years have been marked as a 'NewSpace' era, because of the rise of many private entities carrying out space activities. This is fueled by publicity for projects of SpaceX, Blue Origin and OneWeb. SpaceX for example often reached headlines in 2019 with its new Starlink project. The project is planned to be a global satellite network, aimed at providing reliable and affordable broadband internet services, even (and especially) to remote locations. The first launch took place on May 23, 2019 and consisted of 60 Starlink satellites.[1] The following launches took place on November 11, 2019, January 6, January 29 and February 17, 2020 respectively. Each launch consisted of a batch of 60 satellites, which should amount to a

[1] Starlink mission—mission overview, May 2019 (SpaceX). https://www.spacex.com/sites/spacex/files/starlink_mission_press_kit.pdf, accessed 28 February 2020.

M. Morssink (✉)
Agentschap Telecom (Dutch Radio Communications Agency), Groningen, The Netherlands
e-mail: margaux.morssink@gmail.com

© The Editor(s) (if applicable) and The Author(s), under exclusive license to Springer Nature Switzerland AG 2020

A. Froehlich (ed.), *On-Orbit Servicing: Next Generation of Space Activities*, Studies in Space Policy 26, https://doi.org/10.1007/978-3-030-51559-1_8

total of 12.000 satellites in orbit, following SpaceX's initial plans. A similar example can be found in the case of OneWeb, whose mission it is to launch up to 900 satellites to provide internet services worldwide.

These amounts are staggering when one realizes that up until now, around 9000 space objects are registered. The United Nations Office for Outer Space Affairs (hereinafter: 'UNOOSA') keeps the registry of space objects that have been launched into space. According to its numbers, over 88% of all satellites (and also probes, landers, crewed spacecraft and space station flight elements) launched into orbit have been registered with UNOOSA. The registry contains around 9000 space objects at this moment.[2] Combined with the numbers provided by the European Space Agency (hereinafter: 'ESA'), one could conclude that space is increasingly becoming congested. According to these numbers, since the start of the space age around 5560 rocket launches have taken place. This excludes launch failures. These 5560 rocket launches together have launched around 9600 satellites into space. Of these 9600 satellites, around 5600 are still in orbit.[3] In addition, the annual predictions show that the increase in launches is a trend that will not slow down in the coming years.[4] The prediction for the next years for example is that in 2020, up to 330 small satellites will be launched. For the years 2021–2023, the prediction is that almost 1400 small satellites will be launched. To compare this to the current situation, this is 25% of the satellites that are in orbit now.

Such an increase in satellite launches and the rise of private actors in the space industry raise questions on the sustainable use of outer space, space traffic management and mitigation of space objects and debris. The increase in satellite launches is intrinsically linked to the private sector, which is primarily driven by economic incentives. Because there is a high return on investment, there does not seem to be an incentive to 'clean up' space. On-orbit servicing missions can significantly extend the lifetime of a satellite, so this could be interesting for the parties involved. Keeping a satellite in-orbit for a longer period could be more cost-efficient than building and launching a completely new satellite.

A possible reason for the lack of incentive can be found in the international legal framework, as there are no strict obligations on the private sector in terms of sustainable use of outer space. Because the framework of the United Nations Space Treaties was established in the early stages of space exploration, many technological developments were not taken into account during the draft of the treaties. The Outer Space Treaty and the other Space Treaties provide for a robust legal framework,

---

[2]UNOOSA keeps an online registry: https://www.unoosa.org/oosa/osoindex/search-ng.jspx?lf_id=.

[3]ESA updates these numbers regularly. Space debris by the numbers, February 2020 (ESA). https://www.esa.int/Safety_Security/Space_Debris/Space_debris_by_the_numbers, accessed 01 March 2020.

[4]SpaceWorks for example publishes a small satellite forecast on a yearly basis. In 2019, the forecast of the launches of small satellites in the next 5 years were around 2000–2.800 satellites. The forecasts can be found via the SpaceWorks website.

however often not capable of addressing the technological developments and challenges that have characterized the space industry over the past decade. The solution to addressing new (technological) challenges is either soft law, such as adopting guidelines, or States entering into bilateral agreements with other States or private actors.

8.1.1 Sustainability of Outer Space Activities and Current Legal and Policy Initiatives

In terms of soft law related to the sustainable use of outer space, the Inter-Agency Space Debris Coordination Committee (hereinafter: 'IADC') has established space debris mitigation guidelines. These guidelines can be qualified as 'soft law', as they are a non-binding instrument and not codified law.[5] In itself they do not constitute legal norms, but they are an important instrument in the current framework of the international space treaties. From an international law perspective, adaptation of and adherence to these guidelines by many States and private actors could possibly lead, at one point, to conversion into legally binding rules. The guidelines were the foundation of the Space Debris Mitigation Guidelines of the Committee on the Peaceful Uses of Outer Space. On the application of the guidelines it is stated that:

> Member States and international organizations should voluntarily take measures, through national mechanisms or through their own applicable mechanisms, to ensure that these guidelines are implemented, to the greatest extent feasible, through space debris mitigation practices and procedures. These guidelines are applicable to mission planning and the operation of newly designed spacecraft and orbital stages and, if possible, to existing ones. They are not legally binding under international law.[6]

Although the guidelines are not legally binding, the consistent application of the guidelines by Member States and international organizations could eventually lead to a common, accepted practice.

As the adoption of guidelines show, the sustainable use of outer space has been on the agenda of the United Nations for a couple of years. In 2010, the Working Group on the Long-Term Sustainability of Outer Space Activities was established, under the Scientific and Technical Subcommittee of the UN.[7] Because of the importance of ensuring a safe and sustainable use of outer space, it was stated that a working group should be established 'to support the preparation of a report on the long-term sustainability of outer space activities, the examination of measures that could enhance the long-term sustainability of such activities and the preparation of a set of best-practice

---

[5]Malcolm Shaw, *International Law* (Seventh Edition, Cambridge University Press 2017), pp. 87–88.

[6]Inter-Agency Space Debris Coordination Committee (IADC), Space Debris Mitigation Guidelines, which can be consulted via the IADC website.

[7]Report of the Scientific and Technical Subcommittee on its forty-seventh session, UN Doc. A/AC.105/958, 2010, para 181.

guidelines'.[8] Part of the work of the Working Group is aimed at addressing themes such as sustainable use of outer space, space operations and regulatory regimes for those parties participating in outer space activities.

Developing and adopting guidelines is important in creating an incentive for the private sector to engage in activities related to the sustainable use of outer space, and more specific, on-orbit servicing missions.

8.1.2 On-Orbit Servicing

On-orbit servicing missions have gained more attention over the last years. However, they have been undertaken for decades already, with for example manned missions carrying out repairs on space objects in orbit. The concept of on-orbit (satellite) servicing covers a multitude of possible activities. It can mean repairing satellites that are already in orbit, refueling satellites, carrying out maintenance activities, carrying out maneuvers placing a satellite in a different orbit, assembling a satellite, and so on. It calls for collaboration and communication between the satellite in orbit and the vehicle used for the on-orbit servicing mission. The concept of on-orbit servicing is closely related to the sustainable use of outer space, as it is aimed at extending the lifetime of a satellite, in any kind of manner. In terms of sustainable use of outer space, it can provide a solution to the congestion of outer space. It does however also raise questions on liability in case of damages, as on-orbit servicing missions almost always involve close proximity missions. Those close proximity missions are often hazardous activities with a multitude of possible risks to address.

The first on-orbit servicing mission often referred to is the National Aeronautics and Space Administration's (hereinafter: 'NASA') Skylab project. Skylab was the first space station of NASA and the first example of on-orbit servicing, as its crew performed repair missions on the space station.[9] From the launch on May 14, 1973 until the return of the final crew on February 8, 1974, three different crews worked on the space station, often with success.[10] Another striking example is that of the Hubble telescope, which was repaired on orbit.[11] The first servicing mission on the Hubble telescope, STS-61, took place in 1993. This was only three years after Hubble's

[8]Ibid., para 178.

[9]NASA wrote an extensive history on the Skylab project and its crews. Part I—The History of Skylab, 8 November 2003 (NASA). https://www.nasa.gov/missions/shuttle/f_skylab1.html, accessed 01 March 2020.

[10]On the Skylab mission page on NASA's website, NASA states: *'The effectiveness of Skylab crews exceeded expectations, especially in their ability to perform complex repair tasks. They demonstrated excellent mobility, both internal and external to the space station—showing humans to be a positive asset in conducting research from space'.*

[11]D.E. Hastings & C. Joppin, *On-Orbit Upgrade and Repair: The Hubble Space Telescope Example,* in: Journal of Spacecraft and Rockets, 43(3):614–625.

launch.[12] The servicing mission comprised five extra-vehicular activities performed by astronauts, during which replacements were performed, during which, *inter alia*, solar arrays were replaced, the wide field/planetary camera was replaced and the modifications on the magnetic sensing system were installed.[13]

Since these manned missions in lower Earth orbit, other on-orbit servicing missions have taken place. What makes these missions interesting, is the possibility of significantly extending an in-orbit satellite's mission lifetime. Very recently, an on-orbit servicing mission has taken place, extending the lifetime of a satellite with five more years. What made it more interesting, is that it took place in geostationary orbit and that it was a robotic mission. This mission took place in February 2020, with Northrop Grumman's and SpaceLogistics LLC (a subsidiary of Northrop Grumman) succeeding in docking a vehicle to an Intelsat satellite.[14] The mission is aimed at extending the lifetime of the Intelsat 901 satellite, which was launched in 2001 from Kourou with the United States as the State of registry.[15] The space object that was docked to the Intelsat 901 satellite is the Mission Extension Vehicle-1 (hereinafter: 'MEV-1′). Before the planned return to service, MEV-1 will perform on-orbit services on the Intelsat 901 satellite. Part of the contract between Intelsat and Northrop Grumman is the provision that MEV-1 will provide five years of on-orbit servicing, extending the lifetime of the satellite.[16]

On average, geostationary satellites can be said to have a lifetime of about 10–15 years. Taking into account the recent successful mission, the lifetime of satellites can potentially extend with another five years. Such on-orbit servicing missions could prove to be an interesting way forward, considering extending the lifetime of existing satellites rather than launching new satellites. Supporting new on-orbit servicing missions could also, to some extent, offset the rise in satellites launches, or at least provide for a more efficient use of outer space, in terms of space traffic management.

---

[12]D.J. Shayler & D.M. Harland, *The Hubble Space Telescope: From Concept to Success* (Springer 2015), p. 386. See also: NASA Space Shuttle Mission STS-61 press kit (1993).

[13]Ronald L. Newman, *NASA STS-61 Mission Director's Post-Mission Report* (1995).

[14]Companies demonstrate groundbreaking satellite life-extension service, 26 February 2020 (Northrop Grumman). https://news.northropgrumman.com/news/releases/northrop-grumman-successfully-completes-historic-first-docking-of-mission-extension-vehicle-with-intelsat-901-sat ellite, accessed 01 March 2020.

[15]UN Doc. ST/SG/SER.E/688 (22 October 2013).

[16]See Northrop Grumman's press release, *supra* n. 14.

8.2 Regulation and Legal Framework

On-orbit servicing missions also come with regulatory and legal challenges. Not only does the international legal framework, consisting of the United Nations Space Treaties but also the International Telecommunications Union Radio Regulations, need to be taken into account, there are also often agreements between operators in place, which could prove to be a complicating factor. As there are potentially a couple of parties involved or related to one satellite, it can be difficult to figure out the terms of an on-orbit servicing mission. The parties that could be involved regarding one satellite could be the launching State, the State of registry, the operator and the owner of the satellite. Which rules are applicable to these different entities?

8.2.1 Launching State

The term launching State is the key to impose liability on a State that benefitted from the launch.[17] The term is used in article VII of the Treaty on the Principles Governing the Activities of States in the Exploration and Use of Outer Space, including the Moon and Other Celestial Bodies (hereinafter: 'Outer Space Treaty')[18] which refers to the international liability of a State that launches or procures the launching of an object into outer space. Such States are internationally liable for damage to another State Party to the Outer Space Treaty or to its natural or juridical persons by such object or its component parts on Earth, in air space or in outer space. The liability of States is further elaborated on in the Convention on International Liability for Damage Caused by Space Objects (hereinafter: 'Liability Convention').[19] Following article I of the Liability Convention, the launching State is the State which launches or procures the launching of a space object or a State from whose territory or facility a space object is launched. To summarize, four States can be distinguished as launching States: the State launching a space object, the State procuring the launch, the State from whose territory a space object is launched and the State from whose facility a space object is launched. Any of these four States can be held liable by the State suffering from the damage.[20]

Damage in terms of the Liability Convention can be divided in damage caused on the surface of the Earth or to an aircraft in flight (article II of the Liability Convention) and damage being caused elsewhere than on the surface of the Earth (article III of

[17] L.J. Smith and A. Kerrest, *Article I (Definitions) LIAB*, in S. Hobe, B. Schmidt-Tedd, & K.U. Schrogl (eds.), *Cologne Commentary on Space Law*, vol. 2 (Heymanns 2013), p. 114.

[18] Treaty on Principles Governing the Activities of States in the Exploration and Use of Outer Space, including the Moon and Other Celestial Bodies, adopted on 19 Dec. 1966, entered into force on 10 Oct. 1967, 610 UNTS 205.

[19] Convention on International Liability for Damage Caused by Space Objects, adopted on 29 Nov. 1971, entered into force on 1 Sept. 1972, 961 UNTS 187.

[20] T. Masson-Zwaan & M. Hofmann, *Introduction to Space Law* (Wolters Kluwer 2019), p. 27.

the Liability Convention). In case of damage on the surface of the Earth or to an aircraft in flight, the launching State shall be absolutely liable, whereas in the case of damage caused elsewhere, the principle of fault liability applies. As analysed in the Cologne Commentaries on Space Law, fault is generally seen as 'the failure to adhere to, or breach of, an obligation imposed by law'.[21] As the notion of fault is not seen often in international law and as the Liability Convention does not specify the conditions of fault, in practice it will be difficult to determine fault.[22]

In relation to on-orbit servicing missions, article III of the Liability Convention applies. Damage caused in space falls within the scope of 'damage caused elsewhere than on the surface of the Earth to a space object of one launching State by a space object of another launching State'. Thus, damage caused during an on-orbit servicing mission could result in liability of the launching State if the damage is due to its fault or the fault of persons for whom it is responsible. Apart from determining which State is the launching State (or which are launching States if more States are involved), one also has to determine if the damage is due to the fault of the States involved or persons for whom the States are responsible. This shall be difficult, as the operations take place in orbit. In this sense, it would be a logical step for parties (including launching States) involved in on-orbiting servicing missions to clearly define liability issues in case of damages to the space objects. This applies to launching States involved with the space object that is already in orbit, but also to the launching States of the space object carrying out an on-orbit servicing mission.

8.2.2 State of Registry

The Convention on Registration of Objects Launched into Outer Space (hereinafter: 'the Registration Convention') stipulates the conditions of registration of space objects.[23] As per article I of the Registration Convention, the State of registry is the launching State on whose registry a space object is carried. Article II specifies the condition of registration: when a space object is launched into Earth orbit or beyond, the launching State shall register the space object by means of an entry in an appropriate registry.[24] In case of two or more launching States, these States have to determine which one of them shall register the space object.

Because the State of registry is always (one of) the launching State(s), liability is arranged through the Liability Convention.

---

[21]See *supra* n. 17, p. 132.

[22]See *supra* n. 20, p. 28 and *supra* n. 17, p. 133.

[23]Convention on Registration of Space Objects Launched into Outer Space, adopted on 12 Nov. 1974, entered into force on 15 Sept. 1976, 1023 UNTS 15.

[24]See also *supra* n. 20, pp. 31–32.

8.2.3 Owner of the Space Object

Often, launching States are not related to the owner of the space object. This is because ownership is established before a space object is launched, ‘on the ground’. This is necessary as several agreements need to be made before the space object is launched. Launch service agreements for example regulate the terms of a launch of space object between the launch company and the customer.[25] Those agreements revolve around the obligation to supply the launch, including provisions on *inter alia* launch schedules, replacement launches, waivers of liability. This is all regulated in the sphere of private law and thus the launching State does not have a role in this agreement. The owner of a space object can however play a role in on-orbit servicing missions.

The Outer Space Treaty is directed at States Parties to the Treaty. When it comes to activities of private actors in space, States shall bear international responsibility. This is regulated in article VI of the Outer Space Treaty:

> States Parties to the Treaty shall bear international responsibility for national activities in outer space, including the Moon and other celestial bodies, whether such activities are carried on by governmental agencies or by non-governmental entities, and for assuring that national activities are carried out in conformity with the pro-visions set forth in the present Treaty.

The Outer Space Treaty does however mention ownership of a space object, in article VIII. This can be both private ownership or State ownership[26]:

> Ownership of objects launched into outer space, including objects landed or constructed on a celestial body, and of their component parts, is not affected by their presence in outer space or on a celestial body or by their return to the Earth.

The aim of this article is not to establish ownership. Article VIII stipulates that ownership of a space object established on Earth is not changed when that space object is launched into outer space.[27] That also means that a change of ownership (for example when one private entity sells its satellite to another private entity) does not have an effect on the liability of the involved launching State. This has resulted in provisions in national space laws on authorisation of transfer of ownership from one entity to another. In terms of on-orbit servicing missions, the launching State is liable (if damage caused is due to its fault or the fault of persons for whom the State is responsible, as elaborated on in Sect. 8.2.1). However, when starting an on-orbit servicing mission, the entity that is the owner of the satellite also has to agree with the specific mission. In Sect. 8.1.2 of this chapter an example was mentioned where two entities entered into agreement on an on-orbit servicing mission. The first docking of the vehicle with the satellite has recently taken place. In such a case,

---

[25] See for an extensive summary of contracts in the space sector, including launch service agreements, L. Ravillon, *The Typology of Contracts in the Space Sector*, in I. Baumann & L.J. Smith (edds.), Contracting for Space: Contract Practice in the European Space Sector, pp. 161–168.

[26] B. Schmidt-Tedd and S. Mick, *Article VIII*, in S. Hobe, B. Schmidt-Tedd & K.U. Schrogl (eds.) *Cologne Commentary on Space Law*, vol. 2, p. 163 (Heymanns 2009).

[27] Ibid.

the participants in the mission are limited. This could be different when it comes to government funded missions in terms of a public–private partnership. In such a case, clear arrangements need to be made beforehand with the owners of the space object.

This also applies to another private actor in the chain, the operator of a space object. It could also be possible that a satellite was launched by State A, owned by private entity B and operated by private entity C. The private parties involved in this chain will have contracts in place with waivers of liability, provisions on preventing harmful interference and preventing damage to another space object. In terms of interference, the International Telecommunications Union's Radio Regulations impose obligations on assigning frequencies and preventing harmful interference.[28] As the operator of the space object is also an actor to take into account in relation to on-orbit servicing missions, the ITU Radio Regulations should also be considered. If an on-orbit servicing mission for example does not go according to plan, this could lead to (financial) damage for both the owner and the operator of the space object. This chain of private actors involved in space activities should therefore also be part of an agreement on an on-orbit servicing mission.

8.3 Public–Private Partnerships

To make on-orbit servicing missions more interesting for private actors from an investment and financial perspective, States or government agencies getting involved could provide for an interesting solution. As was elaborated on in Sect. 8.2, the current legal framework comes with challenges as there are several actors involved in on-orbit servicing missions. It is also difficult to impose obligations on private actors in space, as the UN Space Treaties do not specifically address this. Thus the incentive should lie with States, as they can either adopt guidelines (as was addressed in Sect. 8.1.1) or promote specific behavior through adapting their space policy.

Private actors already have the technological and financial means to initiate more sustainable projects in space such as on-orbit servicing missions. Nevertheless, effort from States, especially in terms of legal and policy measures, is needed to make on-orbit servicing more interesting.

The start of creating a different approach to the use of outer space and making on-orbit servicing missions (and as a consequence, extending the lifetime of a satellite rather than launching a new satellite instead) *status quo*, could be entering into public–private partnerships. States or government agencies should initiate these partnerships and thus create an incentive for private actors to take on responsibility and accountability for the activities they undertake in outer space.

---

[28] See for example article 4 of the Radio Regulations on assignment and use of frequencies.

8.3.1 Typology of Public–Private Partnerships

Public–private partnerships are often described as collaborations (partnerships) between the public sector and the private sector for a specific project. Public–private partnerships are often used in the transportation or construction sector but are also applied in relation to activities in outer space. Public–private partnerships can generally be characterized by sharing of risks, sharing of funding and establishing a long-term relationship. In its report on the evolution of the role of space agencies, ESPI has distinguished four common characteristics of public–private partnerships, based on research of the use of these partnerships in practice: 'a long term relationship between public and private partners, funding is sourced from both the public and private sector, an important decisional role is placed on the private entity and partial transfer of risk from public to private parties.'[29]

Such partnerships are beneficial to both parties. For public entities, a public–private partnership can be beneficial because private entities can offer innovative technological ideas and can contribute more financial resources. Projects or entire markets that are too expensive for governments to fully participate in, can be more attractive in the context of a public–private partnership. This is also the case in terms of risk, as risk is shared between the public and private party in the partnership. This is definitely beneficial for the public party, as the private party involved has the most technological knowledge and is best able to analyze possible risks related to the technological activities.[30] For private entities, access can be given to specific markets that were not (easily) accessible before. Partnerships with the public sector can also give financial security (for example, in terms of milestones being met and payment received for meeting these milestones). The private entities engaged in a public–private partnership are also encouraged to, for instance, optimize design and build methods.[31] The public entity is focused on the goal of the public–private partnership, giving the private entity the freedom to execute their part of the partnership. In the design phase, a lot of autonomy is thus given to the private entity. It is free to design the technological part of the project, as it is the party with the most knowledge on this subject. As mentioned above, the duration of the project can give a private party (financial) security. The timeframe related to public–private partnerships can however also be a restraint for private entities. As public–private partnerships are often long term agreements, this can also be considered restrictive for the private party in some projects. Especially in relation to the allocation of risk, as in public–private partnerships the burden of risk is often heavier for the private party.

Because public–private partnerships are beneficial to both parties involved, but can also have their limits, it is important for both parties to establish clear 'rules'

---

[29] Report of the European Space Policy Institute, 'Evolution of the Role of Space Agencies', p. 23 (October 2019).

[30] See *supra* n. 29, p. 66.

[31] See *supra* n. 29, p. 23.

on how the public–private partnership is executed. This is especially the case with public–private partnerships in space, because space activities are hazardous. The risks involved in the partnership therefore need to be assessed as clearly as possible, as was also argued in Sect. 8.2.1 on the allocation of risk in terms of liability. This encompasses the risks from the research and development phase up until the in-orbit phase (and beyond, for example in terms of deorbiting space objects) as there often many factors, both internal and external, that can influence a space project. In this sense, more actors can be involved, possibly making agreement on the terms more difficult.

8.3.2 Public–Private Partnerships in Space

The concept and use of public–private partnerships in outer space is not new. Whereas space activities have traditionally been a prerogative only for States (in the form of their governmental entities), in the decades after the space race, private entities entered the space market. At first, only States had the resources, both financially and technically, to enter into space activities. This is of course primarily due to the space race between the United States and the Russian Federation. As technology advanced, more and more private entities were able to enter the space market as well. Because States needed more funding for their space activities, private entities were able to enter the market and, more importantly, it also became more attractive for these private entities to enter the market.[32]

The past couple of years have been marked as a 'NewSpace' era, because of the rise of many private entities carrying out space activities. In comparison to the early years of space exploration, there is indeed a surge in activities carried out by private entities. However, private entities have been in involved in space activities for quite a while. As a consequence, public–private partnerships related to space activities have existed for quite a while as well.

An example often referred to is the TerraSAR-X mission, which is a collaboration between the German Space Agency (Deutsches Zentrum für Luft- and Raumfahrt, hereinafter: 'DLR') and EADS Astrium GmbH (hereinafter: 'EADS Astrium').[33] EADS Astrium was awarded a contract in 2002 based on a public–private partnership with DLR. The public–private partnership was constructed as follows. It was based on a joint investment by both DLR and EADS Astrium.[34] EADS Astrium, or in fact its company Infoterra, was given the exclusive commercial exploitation rights for the data emanating from the TerraSAR-X remote-sensing satellite. However, the satellite is owned by DLR and the operations of the satellite also remain with DLR.

---

[32] F. Tronchetti, *Fundamentals of Space Law and Policy* (Springer 2013), p. 25.

[33] EADS Astrium was reorganized late 2013/early 2014 into Airbus.

[34] F. Von der Dunk, *European Satellite Earth Observation: Law, Regulations, Policies, Projects and Programmes*, in: 42 Creighton Law Review (2008–2009), pp. 432–433.

The same applies to the scientific data rights, they remain with DLR.[35] The satellite was launched on June 15, 2007,[36] with Germany being registered as its launching State.[37]

Another example of a successful public–private partnership in outer space is the European Data Relay Satellite System (hereinafter: 'EDRS'), which can be dated back to at least 2010, when the tender was issued by ESA. It is part of ESA's ARTES program, which is dedicated to providing the satellite communication industry with the environment to conduct innovative activities and introduce them into the commercial market.[38] ARTES (which stands for Advanced Research in Telecommunications Systems) was established for projects in which the industry can use new and innovative technologies and can also take greater risk, under the wings of the ESA program. In addition, ESA presents the ARTES projects as 'providing maximum benefit with minimal cost and risk to the European taxpayer',[39] which is definitely important being an European space agency. The EDRS project is a collaboration between ESA and Airbus Defence and Space[40] and constructed as a public–private partnerships. It is designed to relay data between satellites in low-Earth orbit, satellites in geostationary orbit and Earth.[41] The main reason for a public–private partnership construction is the goal to achieve a cost-efficient program and to minimize the investments and operation costs on the side of ESA. This is why Airbus Defence and Space acts as prime contractor, building and operating the infrastructure of the system. It is also the owner of the system. ESA funds the research and development part of the mission. As part of the system is used for the Copernicus program, the European Commission is also involved as a customer.[42] The first payload (EDRS-A) and satellite (EDRS-C) were launched in 2015 and 2019 respectively.

A third example of a public–private partnership that might appeal more to the imagination is the National Laboratory on board of the International Space Station (hereinafter: 'ISS'). This is, in contrast to the first two examples, a non-European initiative, as the United States segment of the ISS was designated a United States national laboratory. The designation was enacted through the 'National Aeronautics

---

[35] Ibid. See also the TerraSAR-X Mission page on the ESA website. TSX (TerraSAR-X) Mission (ESA), https://earth.esa.int/web/eoportal/satellite-missions/t/terrasar-x, accessed 28 February 2020.

[36] R. Werninghaus, S. Buckreuss & W. Pitz, *TerraSAR-X mission status*, Proceedings International Geoscience and Remote Sensing Symposium, Barcelona 2007, pp. 3927–3930 (IEEE 2007) and R. Werninghaus & S. Buckreuss, *The TerraSAR-X Mission and System Design*, in: IEEE Transactions on Geoscience and Remote Sensing (2010), vol. 48, no. 2, pp. 606–614.

[37] UN Doc. ST/SG/SER.E/526 (17 December 2007).

[38] Telecom, Artes 4.0 programme—ARTES partnerships projects (ESA), https://artes.esa.int/private-public-partnerships, accessed 28 February 2020.

[39] Applications—Partnership (ESA), https://www.esa.int/Applications/Telecommunications_Integrated_Applications/EDRS/Partnership, accessed 28 February 2020.

[40] Former EADS Astrium, see *supra* n. 33.

[41] C. Al-Ekabi, *European Space Activities in the Global Context*, in C. Al-Ekabi, B. Baranes, P. Hulsroj, A. Lahcen (eds.), Yearbook on Space Policy 2014: The Governance of Space (Springer 2015), p. 29.

[42] See *supra* n. 39.

and Space Administration Authorisation Act of 2005′ (hereinafter: 'NASA Authorisation Act'). The NASA Authorisation Act *inter alia* addresses a commercialisation plan, in which opportunities for the private sector to participate in future missions and activities shall be identified.[43] This includes opportunities for 'partnership between NASA and the private sector in conducting research and the development of technologies and services'.[44] Interestingly, the section on this commercialisation also addresses the need for provisions for developing and funding industry partnerships to conduct commercial research and technology development, to, inter alia, advance the economic interests of the United States. As referred to earlier, the NASA Authorisation Act designated the United States segment of the ISS as a national laboratory. The designation of part of the ISS as a national laboratory, is fueled by the idea of a cost-effective use of the ISS. Increasing the use of the ISS by other federal (United States) entities and the private sector through partnerships, eventually supplements NASA funding of the ISS.

Similar to the goals of ESA's ARTES program, the National Laboratory seeks to open up possibilities for the private sector for innovative research and development projects, as stated on its website: '*This research serves commercial and entrepreneurial needs and other important goals such as the pursuit of new knowledge and education.*' [45]Since the completion of the ISS in 2011, the National Laboratory was used for more than 200 projects, which proves the success of this public–private partnership between NASA and several private entities. Recently, NASA also published a press release on opening the ISS for other commercial business opportunities.[46]

The three examples mentioned in this paragraph are only a part of the public–private partnerships that have been established in relation to outer space activities. What these examples show that in outer space, an area where hazardous activities are conducted, several factors play a large role in the decision whether to conduct these activities or not. For both public and private entities, these factors are, *inter alia*, the allocation or sharing of risk, financial resources, technical capabilities and regulatory boundaries. Both the ARTES/EDRS program and the NASA Authorisation Act are clear on those factors, as the details on the EDRS mission for example explicitly mention that the goal is to achieve a cost-effective program, intended to minimise the investments and operation costs on the side of ESA. A similar goal can be found in the NASA Authorisation Act, which mentions to goal of seeking partnerships supplementing NASA funding of the ISS. This should also not come as a surprise, as the activities conducted in outer space (and prior to being in outer space, in the pre-launch and launch phase) are expensive and government funds are (often) subject

[43]42 USC 16,616. Section 108.

[44]Ibid.

[45]Research on the ISS – Public–private partnerships in space (ISS National Lab), https://www.issnationallab.org/research-on-the-iss/public-private-partnerships-in-space/, accessed 28 February 2020.

[46]NASA Opens International Space Station to New Commercial Opportunities, Private Astronauts, 7 June 2019 (NASA), https://www.nasa.gov/press-release/nasa-opens-international-space-station-to-new-commercial-opportunities-private, accessed 01 March 2020.

to budget cuts. Because of the technological and financial risks related to outer space activities, the concept of a public–private partnership is beneficial to both parties. How public–private partnerships can also accelerate the development of on orbit servicing missions will be addressed in the next section.

8.4 Public–Private Partnerships and On-Orbit Servicing

In Sect. 8.1 the surge in satellite launches and the call for sustainable use of outer space was discussed. Although the rise in launches is primarily linked to the small satellite market in the lower-Earth orbit, questions related to the sustainable use of outer space are not limited to this orbit. For satellites in geostationary orbit, the concept of on-orbit servicing can be very interesting, as it has the possibility of extending the lifetime of a satellite with a significant amount of years. This could offset the rise in satellite launches. The question is why on-orbit servicing missions are not carried out more. It seems that the economic incentive on the side of the private entity plays a large role. In addition, regulatory and legal difficulties could also make it less interesting for a private entity to engage in on-orbit servicing missions. This is where governments can play an interesting role. On-orbit missions have taken place already, but to make on-orbit missions as a service possible, public entities can play an important role as a catalyst.

8.4.1 Why Public–Private Partnerships in On-Orbit Servicing Missions?

As elaborated on in Sect. 8.3, public–private partnerships are characterized by sharing of risks, sharing of funding and establishing a long-term relationship. An important decisional role placed on the private entity is also often part of public–private partnerships. What we have seen so far is that private actors in outer space are mostly driven by financial motives. Generally, there is a high return on investment in a relatively short amount of time. Extending the lifetime of an in-orbit satellite might not be interesting for private actors if it is not seen as beneficial to them. In that sense, public entities should serve as a catalyst to encourage on-orbit servicing missions, because they have the ability to adapt their policy and initiate public–private partnerships. In the next paragraphs, specific characteristics of public–private partnerships are applied to on-orbit servicing missions.

8.4.1.1 Funding

For States or government agencies (hereinafter also referred to as public entities), a public–private partnership can be beneficial because private entities can contribute more financial resources. Government funds are (often) subject to budget cuts. Projects or entire markets that are too expensive for governments to fully participate in, can be more attractive in the context of a public–private partnership. This also applies to on-orbit servicing missions. Because of the high risk involved in on-orbit servicing missions, such missions could also be expensive. This can also be seen in the examples in Sect. 8.3.2, where both ESA and NASA are clear on the financial part of their public–private partnerships. Such arrangements are intended to achieve a cost-effective program and to minimise the investments and operations costs on the side of the public entities. By establishing a public–private partnership programme for on-orbit servicing missions, the burden of costs can be shared between the partners, which will be in favor of the public entity.

In return, engaging in an on-orbit servicing public–private partnership is beneficial for the private entity as it also offers financial security. Apart from the investments that need to be made, a partnership with a public entity, for example to carry out maintenance services to a specific satellite, will generally be for a period of multiple years. This could be a steady element of a private entity's business, if specific milestones of the arrangement are met. In a sector where there are a lot of risks and uncertainties, a public–private partnership could be beneficial for the private entity involved in this sense.

8.4.1.2 Sharing Risks

As was discussed in Sect. 8.2, there are a couple of legal challenges when it comes to on-orbit servicing missions. In potential, six different actors can be involved in an on-orbit servicing mission. The launching State of the in-orbit satellite, the owner of this satellite, the launching State of the space object carrying out the on-orbit service and the owner of this space object. In addition, the operators of the involved space objects could also play a role. From a legal perspective, both the regime of the UN Space Treaties and the ITU Radio Regulations need to be addressed. The on-orbit servicing mission needs to be authorized by a specific State under its national law, conform article VI of the Outer Space Treaty. Avoiding harmful interference, an obligation per the Radio Regulations, also needs to be taken into account.

In public–private partnerships, risks are shared. The risk is often heavier for the private entity. In terms of on-orbit servicing, the launching State bears liability for the damages caused by the space object carrying out the servicing. As public–private partnerships are characterized by shared risks, this should also be the case in on-orbit servicing partnerships. As the private entity carries out the servicing, it shall also have to carry (part of) the burden of risk. This makes a public–private partnership more interesting, especially for the public entity involved. To circumvent issues related to the owner of the satellite involved, one could also choose to limit the public–private on-orbit servicing to satellites owned by space agencies.

8.4.1.3 Technological Knowledge

In public–private partnerships, an important decisional role is often placed on the private entity. Within the terms of the arrangement, a lot of freedom is given to the private entity. It has the most technological knowledge of the two parties and best able to analyze possible risks related to the on-orbit servicing missions. In public–private partnerships in space, private entities are often encouraged to optimize design and build methods. This is why a lot of autonomy is given to the private entity in the design phase. In terms of on-orbit servicing missions, specific attention should be given to technological requirements related to docking to a satellite. Because the possibility of on-orbit servicing perhaps in some cases was not thought of during the design of the satellite, some older satellites might not be designed for docking to a newly designed space object. When it comes to these kind of issues, technological knowledge on the side of the private entity could be very helpful. This also applies to experience with export control issues, which might occur in relation to on-orbit servicing missions.

In this sense, a public–private partnership can be beneficial for both parties involved. Private entities have the technological knowledge but can gain access to markets that were not (easily) accessible before and, for example, make a business model out of on-orbit servicing missions. On the other side, public entities can benefit from the technological knowledge of their partner.

8.4.2 Terms of Public–Private Partnerships in On-Orbit Servicing Missions

A great step forward in stimulating on-orbit servicing missions and cooperation between public and private entities, is the initiative of the Consortium for Execution of Rendezvous and Servicing Operations (hereinafter: 'CONFERS'). CONFERS is an initiative led by the industry, aimed at taking best practices from both the government and the industry to research and develop technical and operations standards for on-orbit servicing.[47] Together, the industry and the government have created standards for on-orbit servicing missions, which is a great step forward in encouraging on-orbit servicing missions. Such initiatives are unequivocally needed to make on-orbit servicing *status quo*.

[47] The Consortium for Execution of Rendezvous and Servicing Operations (CONFERS), https://www.satelliteconfers.org/, accessed 12 March 2020.

Taking into consideration the benefits of public–private partnerships in relation to on-orbit servicing as mentioned in the previous paragraphs, there are also specific terms to think of in relation to these agreements.

First and foremost, because of the close proximity of the two space objects involved in on-orbit servicing, both (or more) parties involved should agree to the mission. This means the owner of the space object, but likely also the launching State(s) involved.

Second and closely related to the first point, liability issues should be addressed beforehand. An on-orbit servicing public–private partnership should have clear arrangements on liability and the shared risks. This is because of the nature of space activities in itself, but even more so because on-orbit services are close proximity activities. This point concerns both the prelaunch phase as the launch and in-orbit phase. Because there are many factors to take into account, the terms of the public–private partnership should be clear on this point.

Third, both parties should agree on ownership of the space object carrying out the servicing mission and the data that emanates from the mission.

In general and perhaps needless to say, public–private on-orbit servicing missions should be carried out in accordance with the same procedures that are applicable to other space activities. This means that they are also subject to licensing and supervision, as per article VI of the Outer Space Treaty. In addition, developing standards for on-orbit servicing missions, as was done by CONFERS, could contribute to public–private on-orbit servicing missions.

8.5 Conclusion

As was set out in the previous paragraphs, to make on-orbit servicing missions the new standard, an impulse from States or their space agencies is needed. At this moment, there is a lack of incentive on the side of the industry, primarily due through the economic benefits that the current business models provide. In relation to the NewSpace developments, some argue that the current international legal framework is outdated and incapable of addressing these new, technological developments. This is to some extent true, as many technological developments were not taken into account during the draft of the UN Space Treaties. However, the current legal framework does not have to change as it provides for a robust legal framework. A lot of the developments States are now confronted with, can be addressed through adopting guidelines as was mentioned in Sect. 8.1.1.

Another way to address the current developments and to benefit from the technological progress that is made by the industry, is to use the knowledge of the industry and to initiate public–private partnerships. Such agreements between States or their agencies and the industry are a great impulse in making on-orbit servicing missions and, as a result, extending the lifetime of in-orbit satellites, the 'new normal'. This 'new normal' is essential as the economic success of private actors in outer space keeps rising and congestion of orbits, space traffic management and the sustainable use of outer space are becoming more pressing issues. In this sense, on-orbit servicing missions and thus extending the lifetime of satellites are required to keep up with the pace of the industry. All actors in space, both public and private entities, should take accountability for the activities they undertake in outer space. Promoting sustainability of outer space is undoubtedly needed to keep facilitating the growth of the industry.

Margaux Morssink graduated from the University of Amsterdam with a Master's Degree in International Law. In addition, she obtained a Master's Degree in Space Activities and Telecommunications Law from the Institut du Droit de l'Espace et des Télécommunications (Université Paris-Sud). She currently works at Agentschap Telecom, the Dutch Radio Communications Agency, at the legal department.

Chapter 9
Legal Aspects Relating to On-Orbit Servicing and Active Debris Removal

Ewan Wright

Abstract Autonomous on-orbit servicing and active debris removal are promising emerging markets in the commercial space industry. However, their advent brings legal challenges that must be overcome for a smooth adoption of the services. This article outlines the history and context of autonomous on-orbit servicing and key legal issues that must be addressed. The lack of clear definition of fault is discussed in the context of close proximity missions, as well as the issue of ownership of debris should a collision occur. Furthermore, the legal implications of the dual use of servicing satellites are discussed. For the commercial market to grow, these legal issues must be addressed through an initial inter-state mission, and the continued development of best practices for on-orbit servicing and active debris removal.

9.1 Introduction

Autonomous on-orbit servicing[1] is a collection of services involving the autonomous rendezvous and docking of two satellites. Satellites docking in space is not novel and the manoeuvre has been conducted many times. However, on-orbit servicing as a commercial service, particularly when the companies involved are from different states, presents some potential legal issues. The first autonomous on-orbit servicing mission is being conducted at the time of writing (Spring 2020), and the market is expected to grow over the coming years with several companies looking to launch their first missions and begin to offer services. Furthermore, a number of active debris removal missions will be conducted in the coming years, presenting similar legal issues to on-orbit servicing.

[1]Occasionally in-orbit is used. The author considers the two synonymous and will use on-orbit for this article.

E. Wright (✉)
University of Sheffield, Sheffield, UK
e-mail: etwright1@sheffield.ac.uk

© The Editor(s) (if applicable) and The Author(s), under exclusive license to Springer Nature Switzerland AG 2020

A. Froehlich (ed.), *On-Orbit Servicing: Next Generation of Space Activities*, Studies in Space Policy 26, https://doi.org/10.1007/978-3-030-51559-1_9

This article does not address the technical or economic challenges faced by satellite servicing missions; it merely discusses legal ramifications of intersatellite servicing and debris removal. It starts with definitions and the history of on-orbit servicing, before discussing the international legal framework and specific issues related to on-orbit servicing and active debris removal.

9.1.1 *Definitions*

On-orbit servicing loosely refers to a collection of services provided by a specially designed servicing satellite rendezvousing with a target satellite. When near the target satellite, the actions taken by the two satellites are referred to as close proximity operations (or rendezvous and proximity operations). The attachment of the servicing satellite with the target satellite is known as docking, with the whole procedure known as rendezvous and docking.

There are several on-orbit servicing missions that have been proposed, with varying technical complexity.

For clarity, the servicing satellite that rendezvouses with the target satellite will be referred to as the approaching satellite. The satellite being serviced or removed from orbit will be referred to as the target satellite.

On-orbit servicing missions that have been proposed include the following:

9.1.1.1 Attitude and/or Orbital Control

The approaching satellite docks with the target satellite and takes over the attitude and orbit control of the target satellite, keeping the target satellite pointing in the right direction and in the correct orbit. The approaching satellite stays docked with the target satellite for the duration of the mission but could go on to service other satellites.[2]

9.1.1.2 Refuelling

The approaching satellite docks with the target satellite and provides it with fuel to extend its mission, before undocking. It is difficult to know how much fuel is left in a spacecraft, so this service would be useful for reducing uncertainty around mission duration and ensuring enough fuel is left to move the satellite to a disposal

[2] Jeff Foust, 'Rethinking Satellite Servicing', *The Space Review*, 4th February 2019. https://www.thespacereview.com/article/3653/1, (accessed 27th March 2020).

orbit.[3] Each year, around ten satellites go out of service due to lack of fuel that could potentially be refuelled if such a service existed.[4]

9.1.1.3 Repair

The servicing satellite manipulates the target satellite to repair it. This could include opening a solar panel that has failed to deploy, or more advanced repairs, including replacing parts.

9.1.1.4 Active Debris Removal

The aim of the mission is to remove a defunct satellite or debris from orbit. The approaching satellite captures the target debris before deorbiting. A variety of capture methods have been proposed, including harpoons, nets, magnets, and mechanic manipulators. Whilst not strictly on-orbit servicing, it is considered here due to the legal similarities between active debris removal and on-orbit servicing.

9.1.2 History

The following section outlines some key milestones in the history of on-orbit servicing and active debris removal.

The first rendezvous and docking procedure in space was achieved in 1966 when Gemini 8, piloted by Neil Armstrong, docked with an uncrewed Agena target vehicle.[5]

In 1984, the Space Shuttle (STS-51A) retrieved Palapa B2 and Westar VI, two communication satellites that were stranded outside of their intended orbit after their apogee kick motors failed.[6] The satellites were not designed to be docked with and

---

[3]Brian Weeden, 'Dealing with Galaxy 15: Zombiesats and on-orbit servicing', *The Space Review*, 24th May 2010. https://thespacereview.com/article/1634/2, (accessed 27th March 2020).

[4]Jeff Foust, 'The space industry grapples with satellite servicing', *The Space Review*, 25th June 2012. https://www.thespacereview.com/article/2108/1, (accessed 27th March 2020).

[5]Barton C. Hacker and James. M. Grimwood, 'On the Shoulders of Titans: A History of Project Gemini', *NASA Special Publication-4203*, 1977. https://www.hq.nasa.gov/office/pao/History/SP-4203/ch13-6.htm, (accessed 29th March 2020).

[6]Jeanne Ryba and Brian Dunbar, 'Mission Archives: STS-51A', *NASA's John F. Kennedy Space Centre Mission Archives*, 18th February 2010. https://www.nasa.gov/mission_pages/shuttle/shuttlemissions/archives/sts-51A.html, (accessed 29th March 2020).

were returned to Earth for repair and relaunch,[7] arguably a form of active debris removal.

The first rendezvous and docking procedure for the purpose of servicing a satellite was the STS-49 mission of 1992, whereby the Space Shuttle crew retrieved the communication satellite Intelsat 603 and attached it to a new upper stage, allowing it to move to geosynchronous orbit.[8]

Other notable missions included the servicing of the Hubble Space Telescope (HST). This comprised five missions conducted between 1993 and 2009 where the Space Shuttle rendezvoused and docked with the HST. Astronauts performed repairs and upgrades to the satellite by hand, including repairing gyroscopes and installing new solar panels and scientific instruments.[9]

In 2007, as part of the Orbital Express programme, two satellites, ASTRO and NEXTSat, rendezvoused and docked autonomously in space using a mechanical arm, transferring fuel and testing other robotic servicing technologies.[10]

In early 2020, Northrop Grumman subsidiary Space Logistics' Mission Extension Vehicle (MEV) docked with Intelsat-901, the first autonomous docking with a spacecraft not designed to be docked with. The docking of the MEV-1 mission occurred in geostationary graveyard orbit, away from active satellites, to minimise risk to other spacecraft if the mission did not succeed.[11] The second MEV mission, MEV-2, is planned for later in 2020.[12]

9.1.3 The Future

Several factors have coincided that mean on-orbit servicing is an emerging market:

---

[7]Richard Parker, 'On-orbit satellite servicing, insurance and lessons of Palapa B2 and Westar 6' *ROOM Issue #1(3)*, 2015 https://room.eu.com/article/Onorbit_satellite_servicing_insurance_and_lessons_of_Palapa_B2_and_Westar_6, (accessed 27th March 2020).

[8]Jeanne Ryba and Brian Dunbar, 'Mission Archives: STS-49', *NASA's John F. Kennedy Space Centre Mission Archives,* 31st March 2010. https://www.nasa.gov/mission_pages/shuttle/shuttlemissions/archives/sts-49.html, (accessed 30th March 2020).

[9]Rob Garner and Brain Dunbar, 'About—Hubble Servicing Missions', *NASA Mission Pages,* 8th April 2020. https://www.nasa.gov/mission_pages/hubble/servicing/index.html, (accessed 30th March 2020).

[10]Brian Berger, 'U.S. Air Force to End Orbital Express Mission', *Space.com,* 20th June 2007. https://www.space.com/4018-air-force-orbital-express-mission.html, (accessed 30th March 2020).

[11]Caleb Henry, 'Northrop Grumman's MEV-1 servicer docks with Intelsat satellite', *SpaceNews.com,* 26th February 2020. https://spacenews.com/northrop-grummans-mev-1-servicer-docks-with-intelsat-satellite/, (accessed 27th March 2020).

[12]Michael Sheetz, 'For the first time ever, a robotic spacecraft caught an old satellite and extended its life', *CNBC,* 17th April 2020. https://www.cnbc.com/2020/04/17/northrop-grumman-mev-1-spacecraft-services-intelsat-901-satellite.html, (accessed 3rd April 2020).

- The retirement of the Space Shuttle has meant that there is no spacecraft available to repair satellites. Satellites that have failed and could have been repaired by the Space Shuttle are now left stranded on-orbit.
- There is significant uncertainty in the geostationary satellite communications sector as revenues from broadcast decline, and new satellite technologies such as low Earth orbit (LEO) constellations and high throughput broadband satellites launch. This has reduced the number of geostationary satellites ordered each year from 20–25 to 7–8 as operators delay ordering new satellites. This has strengthened the business case for servicing: satellite communication companies may choose to pay in the order of tens of millions of dollars to extend the revenues of an existing satellite rather than ordering a new satellite for hundreds of millions of dollars. In the case of Intelsat-901, the 19-year old satellite will go on to operate for five more years.
- New constellation operators need to ensure their orbits are kept as clear as possible and may pay for an active debris removal service. More stringent national regulations regarding orbital debris may require this. Currently only about 20% of satellites above 650 km attempt to deorbit.[13] For these satellites, where the satellite would not deorbit naturally, active debris removal will be essential to remove the satellites from orbit.

The Northrop Grumman MEV-1 mission is the first where a company is providing a commercial service in extending the life of the satellite. Over the next decade, this is expected to become a growing market, with Northern Sky Research, a market research consultancy, estimating cumulative revenues worth $3.1 B up to 2029.[14] Northrop Grumman expect to launch further, more advanced missions, joined by start-ups including Effective Space[15] and Infinite Orbits[16].

Within active debris removal, start-up Astroscale has raised $153 m[17] and hopes to launch demonstrator mission in 2020,[18] and ClearSpace has agreed a deal worth

[13] Jeff Foust, 'Mega-constellations and mega-debris', *The Space Review*, 10th October 2016. https://www.thespacereview.com/article/3078/1, (accessed 3rd April 2020).

[14] Northern Sky Research, 'In-orbit satellite services pave the way to manage space assets', *Northern Sky Research.* 11th February 2020. https://www.nsr.com/nsr-report-in-orbit-satellite-services-pave-the-way-to-manage-space-assets/, (accessed 30th March 2020).

[15] Effective Space, 'Effective Space Fact Sheet'. https://www.effective.space/, (accessed 30th March 2020).

[16] Infinite Orbits, 'About IO'. https://www.infiniteorbits.io/, (accessed 30th March 2020).

[17] Crunchbase, 'Astroscale'. https://www.crunchbase.com/organization/astroscale#section-overview, (accessed 5th April 2020).

[18] Jason Forshaw and Andy Bradford, 'The ELSA-d End-of-life Debris Removal Mission: Preparing for Launch' *70th International Astronautical Congress (IAC)*, 21st October 2019, IAC-19,A6,5,2,x49982. https://astroscale.com/wp-content/uploads/2020/02/ELSA-IV-Conference-IAC-2019-v1.1.pdf, (accessed 29th March 2020).

$100 m from the European Space Agency to demonstrate an active debris removal mission in 2025.[19]

9.1.4 *Tracking Considerations*

Tracking of satellites in orbit is mostly performed by governmental organisations, notably the US Joint Space Operations Command, who track over 20,000 space objects larger than 10 cm across and share this information with operators around the world.[20]

Of these tracked space objects, around 2200 are satellites.[21] 562 operate in geostationary Earth orbit (GEO), where the tracking is less accurate due to the larger distance from the ground sensors.[22]

Detection of objects in orbit is performed by a range of radars and telescopes across the globe. These data are then combined to create a model of the satellites path, which is propagated into the future to estimate the satellites trajectory. This provides an estimate of where a satellite will be, and whether it will conflict with another satellite. However, there is uncertainty in these estimated paths, and operators cannot be exactly sure of the future path of their satellite.

9.2 Legal Considerations

This section will outline the legal framework of international space law before addressing key issues related to on-orbit servicing.

The primary body of international space law consists of five international treaties. The first two treaties, the Treaty on Principles Governing the Activities of States in the Exploration and Use of Outer Space, including the Moon and Other Celestial Bodies (hereinafter Outer Space Treaty) of 1967 and the Convention on International Liability for Damage Caused by Space Objects (hereinafter Liability Convention) of 1972 are most relevant to on-orbit servicing. The Outer Space Treaty, ratified by 109 states, outlines general principles of conduct in outer space, and the Liability

[19]European Space Agency, 'ESA commissions world's first space debris removal', *ESA Safety and Security*, 9th December 2019. https://www.esa.int/Safety_Security/Clean_Space/ESA_commissions_world_s_first_space_debris_removal, (accessed 30th March 2020).

[20]T. A. Aadithya, 'Review Paper on Orbital Debris Mitigation and Removal and a New Model Insight', *Proceedings of the 24th IRF International Conference*, 3rd May 2015, ISBN: 978-93-85465-07-9.

[21]Union of Concerned Scientists, 'UCS Satellite Database December 2019 Update', 16th December 2019. https://www.ucsusa.org/resources/satellite-database, (accessed 30th March 2020).

[22]Brian Weeden, 'Dancing in the dark redux: Recent Russian rendezvous and proximity operations in space', *The Space Review*, 5th October 2015. https://www.thespacereview.com/article/2839/2, (accessed 29th March 2020).

Convention, ratified by 96 states, outlines the liability for damage caused by space objects on the earth and in space.

9.2.1 Outer Space Treaty

The primary legal issue raised by on-orbit servicing is the consequences of a collision between the two satellites if they launched from different states.

Article VII of the Outer Space Treaty states:

> Each state (...) is internationally liable for damage to another State Party to the Treaty or its natural or juridical persons by such object or its component parts on the Earth, in air space or in outer space.[23]

For the case of on-orbit servicing, states would be liable for damage caused if their satellite damaged a satellite from another state. It is not clear which state would be liable for damage; there is no test for liability. This is elaborated upon in the Liability Convention.

Article IX of the Outer Space Treaty states:

> If a State Party to the Treaty has reason to believe that an activity or experiment planned by it or its nationals in outer space, (...), would cause potentially harmful interference with activities of other States Parties in the peaceful exploration and use of outer space, (...), it shall undertake appropriate international consultations before proceeding with any such activity or experiment.

This implies that if satellite servicing missions are to be conducted between states, including active debris removal, states should reach some sort of legal agreement beforehand. Previously, autonomous satellite dockings have been conducted intra-state, so there has been no international law requirement for an agreement outlining liability if something went wrong, though commercial entities may have created them anyway.

The Outer Space Treaty uses vague language[24] and was written at a time when preventing a war in space took priority. It could not have predicted the advent of autonomous on-orbit servicing over 50 years later. The Liability Convention clarifies some points.

[23]United Nations Office for Outer Space Affairs, *Treaty on Principles Governing the Activities of States in the Exploration and Use of Outer Space, Including the Moon and Other Celestial Bodies*, 1966.

[24]Cristin Finnigan, 'Why the Outer Space Treaty remains valid and relevant in the modern world', *The Space Review*, 12th March 2018. https://www.thespacereview.com/article/3448/1, (accessed 5th April 2020).

9.2.2 Liability Convention

Elaborating upon Article VII of the Outer Space Treaty, the Liability Convention introduces fault-based liability in the case of in-orbit collisions. Article III of the Liability Convention states:

> In the event of damage being caused elsewhere than on the surface of the earth to a space object of one launching state or to persons on board such a space object by a space object of another launching state, the latter shall be liable only if the damage is due to its fault or the fault of persons for whom it is responsible.[25]

However, the Liability Convention does not define fault. Furthermore, it has not been formally invoked, so it is yet to be tested.[26]

Black's law dictionary defines fault as:

> 1. An Error or defect of judgement or conduct; any deviation from prudence or duty resulting from inattention, incapacity, perseverity, bad faith, or mismanagement (…) 2. The intentional or negligent failure to maintain some standard of conduct that failure results in harm to another person.[27]

This definition leads to some key questions relating to on-orbit servicing:

Is the approaching satellite at fault, because it approached the target satellite, or is the target satellite at fault for being close to failure in the first place? Which satellite malfunctioned at the critical time, causing the collision? Did the satellites make contact before the breakup? These are scenarios that must be considered in order to prepare legal arrangements for servicing missions.

9.2.3 Debris Creation

Article I of the Convention on Registration of Objects Launched into Outer Space (hereinafter Registration Convention) of 1974, ratified by 69 states, defines a space object as "component parts of a space object as well as its launch vehicle and parts thereof".[28]

There is no legal definition for space debris. The UN Committee On the Peaceful Uses of Outer Space Space Debris Mitigation Guidelines of 2007, endorsed by the General Assembly, define space debris as "all man-made objects including fragments

---

[25]United Nations Office for Outer Space Affairs, *Convention on the International Liability for Damaged Caused by Space Objects,* 1971.

[26]Secure World Foundation '2009 Iridium-Cosmos Collision Fact Sheet', *Secure World Foundation,* 10th November 2010. https://swfound.org/media/6575/swf_iridium_cosmos_collision_fact_sheet_updated_2012.pdf, (accessed 5th April 2020).

[27]*Black's Law Dictionary,* 8th ed. 2004, 641.

[28]United Nations Office for Outer Space Affairs, *Convention on Registration of Objects Launched into Outer Space,* 1974.

and elements thereof, in Earth orbit or re-entering the atmosphere, that are non-functional."[29]

States are liable for objects they launch into space. This also includes any debris created through the duration of a mission, or during a collision or breakup. Theoretically, states are responsible for even a fleck of paint that comes off their satellite, though this is moot due to smaller objects not being trackable. If a piece of debris went on to damage another states' satellite, the state that caused the debris would be liable.

When docking two satellites, accidents can occur. This will be particularly relevant over the next decade as proof of concept missions are conducted, including by start-ups conducting their first space missions. The risk of largest consequence is that there is a catastrophic collision causing creation of a large amount of space debris. A smaller consequence could be failure of either satellite, or more minor damage. It is not uncommon in close proximity operations for the two satellites gently collide, or 'bump', with no debris creation or notable damage.[30]

Under the strict liability regime above, debris remains the responsibility of the launching state. The two states involved in a collision would be liable for their own debris if it went on to damage other states' space objects. In a collision, especially when the angle of inclination between the satellites is low, we cannot reliably determine which debris comes from which spacecraft. Instead, the state at fault could be liable for all debris created in the collision. Determining which party is responsible for the crash is therefore paramount.

Claims from collisions caused by resulting debris may not be viable, but they could be credible enough to create international tension. Responsibility for ensuing debris from on-orbit servicing accidents would mean states would have an even stronger incentive to avoid collisions.

9.2.4 Active Debris Removal

In the case of active debris removal, the target satellite is non-functioning and therefore defined as debris. It could be argued that a collision is almost certainly the approaching satellite's fault. However, if this has significant consequences for the approaching satellite operator it could be hindering to the greater mission of reducing space debris. Clearly, some pragmatic solution is required to make liability arrangements workable.

[29]United Nations Office for Outer Space Affairs, 'Space Debris Mitigation Guidelines of the Committee on the Peaceful Uses of Outer Space', *UNOOSA*, January 2010. https://www.unoosa.org/pdf/publications/st_space_49E.pdf, (accessed 30th March 2020).

[30]Brian Weeden, 'Dancing in the dark redux: Recent Russian rendezvous and proximity operations in space', *The Space Review*, 5th October 2015. https://www.thespacereview.com/article/2839/2, (accessed 29th March 2020).

One solution could be to adopt the common law doctrine of abandonment for space debris.[31] This would involve treating non-functioning space objects as abandoned. Objects classed as abandoned would not require a formal transfer of ownership or international legal agreement in order to be removed by another state. Furthermore, it would reduce the strict ownership regime imposed by the more stringent readings of the definition of a state's space object. However, the issue of liability still remains, as abandoned space objects could still go on to damage spacecraft and a state must still be liable. Moreover, reducing a state's responsibility for its debris would reduce the incentive to remove its own debris.

A formally agreed permission to engage framework[32] could be involved in allowing states to approach and remove space debris. This could involve consultations between the UN and states to allow states to claim space objects and, if no response received, clearing missions could then remove the debris with some degree of confidence that they are not violating the Outer Space Treaty. It could also allow for space objects to be declared abandoned, as above, with the UN.

9.2.5 On-Orbit Servicing Satellites as Weapons

A consideration of servicing satellites is that they could also be used by hostile actors to bring down functioning satellites.[33] This is accurate, and a satellite could rendezvous with another satellite and either damage it or deorbit it. Due to tracking from the ground, the target satellite would likely have some warning, and would likely know which state the hostile spacecraft came from by either registration or records of launch.

Nearly all satellites are dual use, with both military and civil applications. Satellites could potentially be controlled to collide with other satellites, though this would be difficult without rendezvous and proximity capabilities. However, using a servicing satellite may not be the most effective method of disabling satellites. Missiles could be used to shoot down satellites and ground and space-based jamming signals could disrupt control to satellites. Using a servicing satellite to attack an adversary's satellite would reduce debris, which is admirable but likely not the hostile actor's highest priority.

---

[31] Chelsea Muñoz-Patchen, 'Regulating the Space Commons: Treating Space Debris as Abandoned Property in Violation of the outer Space Treaty', *Chicago Journal of International Law,* 16th August 2018, Vol. 19: No. 1, Article 7.

[32] Brian Weeden, 'Overcoming Legal, Policy, and Economic Hurdles for Active Debris Removal', *OECD Workshop on Economics of Space Debris,* 19th June 2019. https://swfound.org/media/206465/bw_oecd_overcoming_non-technical_challenges_adr_june2019.pdf, (accessed 30th March 2020).

[33] Jeff Foust, 'The space industry grapples with satellite servicing', *The Space Review,* 25th June 2012. https://www.thespacereview.com/article/2108/1, (accessed 27th March 2020).

From a legal perspective, the use of a hostile servicing satellite to disable or damage another state's satellite would be an attack and the body of international humanitarian law applies. It would also be contrary to the 'peaceful purposes' restriction of the Outer Space Treaty. Reponses should be guided by the customary international law of self-defence principles of necessity and proportionality.

However, if states develop the technologies for on-orbit servicing in a secretive way, political tensions between states could increase.

9.2.6 International Test Mission and International Customary Law

A test mission between two organisations registered in different states would provide some precedent and resolution for these legal issues.[34] Companies and states engaged in on-orbit servicing should make clear their arrangements regarding fault and the ownership of debris should an accident happen. This would help develop best practices around on-orbit servicing and inter-state rendezvous and docking.

Continued missions between states could create evidence of *opinio juris* across states, allowing the creation of customary international law and binding states to best practices.

9.3 Conclusion

States have conducted rendezvous and docking procedures for over 50 years. More recently, commercial demand has increased the possibility of using rendezvous and docking technology to service satellites autonomously or remove them from orbit. When conducted by organisations in different states, this raises a number of legal issues. The main issue is the ownership of any debris created and liability for fault should an accident occur. For companies to offer services, these legal issues must be addressed. As companies conduct their first missions and begin to offer commercial services, an international test mission would provide an opportunity to resolve these issues and ensure this emerging market is able to flourish.

Ewan Wright is currently studying for a Master's degree in Aerospace Engineering at the University of Sheffield, having spent a year working as a business analyst in the commercial space industry. He is particularly interested in the commercial adoption of emerging space technologies and the ramifications of the growth in space debris.

[34] Brian Weeden, 'How Do I Ask Permission to Engage With A Piece of Space Debris', *3rd European Workshop on Space Debris Modelling and Remediation,* 16th June 2014. https://swfound.org/media/171984/weeden_permission_to_engage_june2014.pdf, (accessed 29th March 2020).

The manufacturer's authorised representative in the EU is Springer Nature Customer Service Centre GmbH, Europaplatz 3, 69115 Heidelberg, Germany. If you have any concerns regarding our products, please contact ProductSafety@springernature.com

Printed and bound by CPI Group (UK) Ltd, Croydon, CR0 4YY
30/09/2026
02225515-0014